Neabsco and Occoquan: The Tayloe Family Iron Plantations, 1730–1830

Neabsco and Occoquan: The Tayloe Family Iron Plantations, 1730–1830

Laura Croghan Kamoie

Prince William Historical Commission

ISBN 0-9747251-0-2

Library of Congress Control Number: 2003114777

Table of Contents

Tables

Illustrations

Introduction

Dr. Laura Croghan Kamoie began her research on the Tayloe iron plantations while she was an advanced degree student at the College of William and Mary. Her intention was to research this prominent family's entrepreneurial exploits throughout as part of her doctoral dissertation, "Three Generations of Planter-Businessmen: The Tayloes, Slave Labor, and Entrepreneurialism in Virginia, 1700–1830." Under the guidance of Dr. Lawson Pendleton, a former member of the Prince William County Historical Commission, Laura was persuaded to allow the county to publish the part of her dissertation that focused on the Tayloe family's presence in our county.

What her research has brought to light not only elucidates the commercial ventures of the Tayloe family but properly places Prince William County as an important link in those ventures. We learn from Dr. Kamoie's essay the variety of ways the enterprising Tayloe family used the natural resources of the county to build wealth, support the American Revolution, expand the mail routes, and provide transportation and shelter for those who traveled its roads.

Dr. Kamoie has also given back to us the names of some of the earliest residents of the county. We learn of Billy, the runaway slave, Travers Nash, the head clerk of the ironworks, and John Ballandine, the troublesome partner of John Tayloe II in the Occoquan Ironworks. It was the labor of such men and women—enslaved, indentured, and free—that helped lay the foundation for the Tayloes' commercial successes and their wealth that would eventually enable them to build grand establishments not only in Virginia but in the nation's center of power, Washington, D.C.

The Prince William County Historical Commission is very grateful to Laura Kamoie for her extensive research. We thank her for writing the essay *Neabsco and Occoquan: The Tayloe Family Iron Plantations, 1730–1830* and allowing the Historical Commission to publish it so that current residents might have the opportunity to learn of their county's past.

Mary Altomare Raimondi
Prince William County Historical Commission

Neabsco and Occoquan:
The Tayloe Family Iron Plantations, 1730–1830

For over a century, the Tayloe family played an important role in the development of Prince William County, Virginia. Over the course of the century between 1730 and 1830, three generations of this family built communities, founded businesses, and sailed the waterways of Prince William County. The family's formal seat was in Richmond County in the Northern Neck of Virginia. But the family's industrial and transportation enterprises—two legs of the Tayloe business empire—were situated almost entirely in Prince William. At the height of their involvement in the county, just before the death of John Tayloe II in 1779, the family owned approximately 20,000 acres—or over 30 square miles—of land roughly bounded by the Occoquan River on the north, the Potomac River on the east, Quantico Creek on the south, and extending inland to the west for several miles past present-day Interstate 95. During the eighteenth century, the Tayloes were instrumental in the development of Occoquan and Dumfries, and were renowned for their ironmaking and shipbuilding enterprises at the Neabsco and Occoquan Ironworks. These activities were part of the Tayloe patriarchs' efforts to secure the family's fortune and social position in Virginia, and were similar to efforts that elite planters throughout the Chesapeake were making to protect themselves from fluctuations in the tobacco economy.

Beginning during the economic recession of 1680–1720 and continuing through the eighteenth century, Chesapeake planters regularly attempted to find new business activities and investments to battle diminishing tobacco returns.[1] The patriarchs of the Tayloe family of Richmond County, Virginia—John I (1687–1747), John II (1721–1779), and John III (1771–1828)—all invested in, established, and operated alternative economic activities in order to supply themselves with less expensive locally produced consumer goods and establish a buffer against the worst fluctuations of the tobacco market.[2] Diversification was more than a reaction by planters to downturns in the market; it was a long-term trend that shaped local exchange, social interaction, planters' labor requirements, the nature of servants' and slaves' work, and planter class identity.[3] The wealthiest planters possessed the greatest resources in land and slaves and thus moved earliest and most consistently into diversification. Many of these planters had always looked for new and improved ways of making money.[4] These men were not just planters; they were planter-businessmen with a two-fold commitment to both agriculture and a variety of non-agricultural business enterprises. The Tayloes were leaders in one industry in particular—ironmaking. The iron industry was one of the riskiest yet most lucrative businesses in the colonial Chesapeake. In the century between the 1720s and 1830s, the Tayloes founded and operated five major iron plantations in Virginia, of which the Neabsco Ironworks in Prince William County was the most important and enduring.

John Tayloe I and the Bristol Ironworks

John Tayloe I was the founder of the family's iron legacy. A second-generation Virginian, upon adulthood Tayloe quickly acquired political power and the reputation as a businessman.[5] It appears that this Tayloe became involved with the iron industry through a partnership with some of his merchant associates in England, who around 1720 began developing a plan to establish an ironworks in Virginia. Merchants John King, Jeremy Innys, John Lewis, Samuel Jacob, Lyonel Lyde, Walter King,

Figure 1
Potomac River Landings

The majority of the Tayloe family's property was located in the Potomac Region in the Northern Neck of Virginia, northern Virginia, and southern Maryland. Mt. Airy Plantation, the main family seat, was located in Richmond County. The Neabsco and Occoquan Ironworks were situated northwest of Mt. Airy in Prince William County. (Courtesy of the Library of Virginia)

John Templeman, and Samuel Dyke were the core group of Virginia and Bristol investors who founded a company called the Bristol Ironworks in early 1721. The merchant-investors' other interests prevented them from overseeing the day-to-day operations of the works. Thus, they appointed several long-time business associates, including John Tayloe, John Lomax, and Philip Elway, as partners and managers of the ironworks. Tayloe seems to have performed the lion's share of the managerial duties, as by 1724 the merchants appointed him sole "Principal agent overseer and Director of the said Iron works." His duties included supervising workers and production, purchasing and selling tracts of land, and acquiring the necessary raw materials, including iron ore and wood for charcoal.[6]

Tayloe and the Bristol merchants were not alone in realizing the lucrative opportunities of the iron industry. During the seventeenth century, England had limited colonial iron production to prevent competition with her own furnaces.[7] But by the 1710s, England was importing more iron from Sweden than she was producing herself. When King George I restricted the Swedish trade with an embargo in 1717, England looked to the colonies to fill the gap.[8] Virginians were then able to use England's deficiency to their advantage, and induced England to finally allow the production and exportation of iron in the 1727 "Act for encouraging Adventurers in Iron-Works," which provided for internal improvements and tax exemptions for "persons responsible for running or working iron works."[9] However, Chesapeake planter-businessmen did not wait for this official encouragement to begin establishing ironworks. Alexander Spotswood established the Tubal Furnace in Spotsylvania County by 1719. Spotswood also founded three other furnaces, Massaponax Air, Fredericksville, and Germanna, in 1727, 1728, and 1732 respectively.[10] A group of British merchants, in an arrangement similar to the one the Bristol merchants would use a year later, established the Principio Company in 1720. The Pine Forge was founded in the Shenandoah region in 1725, and the Baltimore Company began operations by 1731. The successes of larger establishments such as the Principio and Baltimore Companies, both in Maryland, spurred others, such as Tayloe, John Snowden, and William Byrd II, to erect their own ironworks during the 1730s. Philemon Lloyd, an important Maryland planter and official, described the situation best in 1722 when he said that the iron industry was spreading "more or less all over the Country."[11]

Once appointed, Tayloe wasted no time in getting the new works underway. Within months, Tayloe chose a location on a branch of the Rappahannock River in King George County for the furnace. He purchased the initial lands on what became known as the Bristol Mine Run in 1721 and regularly added other lands over the next two decades.[12] Altogether, the Bristol Ironworks Company owned over 1,350 acres by the 1730s. Such an extensive estate was necessary for the successful operating of an ironworks, as furnaces required huge amounts of raw materials.

While it is nearly impossible to determine the production and profit levels of the Bristol Ironworks, the Bristol merchants expressed their support of Tayloe's management over the years. For example, in 1724, Tayloe was appointed the sole director of the works. Then, in 1728, for "Divers and Manifold good Services. . . . and also for the respect and esteem they . . . the said partners beareth toward the said John Tayloe" the partners offered Tayloe all the remaining available shares in the company amounting to a 10 percent interest. The following year the partners reaffirmed Tayloe, their "friend and partner," as the power of attorney for all matters related to the works.[13] During the 1730s, the partners seem to have given Tayloe a free hand at expanding and shaping the works as he saw fit.

By the 1740s, however, the Bristol partners had apparently lost interest in the Virginia works, and authorized Tayloe to "Dispose of all" of their interests in the property to any interested purchaser. In March of 1741, Tayloe sold the main part of the furnace estate containing 440 acres to a partnership including Ralph Falkner and Edward Neale of Maryland, and Virginians John Triplett, Charles Ewell, and Nathaniel Chapman for £225 sterling.[14] Tayloe handled the "B Co. Disburstments [sic]," or the partners' shares of the money made from the sale of the estate. However, Tayloe appears to have stayed involved with the Bristol works, if only briefly. In 1741, he received an annual salary of £60, and recorded having paid £10 for "the Rent of the Furnace 1 Years." It is possible that Tayloe continued to operate the works on behalf of himself, as he owned a tenth of the original company, and the new partner-owners. Tayloe likely continued to manufacture iron at Bristol through at least 1742. Indeed, in 1741, the works received 903 tons of iron ore, enough to keep the furnace in blast for almost 300 continuous days.[15] In 1742, however, Tayloe sold what remaining company property he owned, probably signaling the end of his involvement with as well as the closing of the Bristol Ironworks.[16]

The end of Bristol did not mark the end of Tayloe's involvement with the colonial iron industry. Instead, it marked the beginning of an enterprise in which three generations of the Tayloe family would participate. As the 1730s wore on, Tayloe apparently developed a desire to work for himself. Indeed, it is likely that even as early as January of 1728/9, Tayloe was looking for a new venture in which to engage. At that time, he wrote to Robert "King" Carter, friend and fellow businessman, with an offer. Knowing that Carter had interests in copper mining, Tayloe urged him to join him in an iron-mining venture. Carter declined, replying, "At this time I am [not] willing to be a joynt Undertaker with your Society untill I am better acquainted with the progress you are likely to make." Carter had learned the hard way to choose his investments carefully. The previous year, his Frying Pan Copper Mine Company in present-day Fairfax County folded when British merchants found the quality of Carter's ore mediocre. Yet, Carter encouraged Tayloe. "I am glad to find so many considerable gentlemen are engag'd in designs of digging Treasure out of the Earth and do heartily wish them success in their Undertakings."[17]

The Founding of the Neabsco Ironworks

These initial plans for a new venture came to fruition by 1737. In that year, Tayloe founded his own ironworks on the Neabsco Creek, a branch of the Occoquan River, in Prince William County. During the 1730s, Tayloe had steadily accumulated large tracts of land in Prince William. By 1737, he owned approximately 9,000 acres bound on the south by the Quantico Creek, on the north by the Occoquan River, and on the east by the Potomac River.[18] One of Tayloe's first concerns, and one he had discussed with Carter, was the issue of obtaining the necessary iron ore for production. Since Tayloe's lands were part of the "Fairfax Proprietary," a tract of over 5,000,000 acres in northern Virginia granted to the Honorable Thomas Lord Fairfax Baron Cameron, Tayloe would be required to pay special fees and taxes on one-third of all ores and minerals his lands produced.[19] While his Neabsco lands did contain some ore deposits that he would use from time to time, Tayloe decided to tackle the problem by finding a source not subject to the proprietary rules.

In 1738, Tayloe purchased lands containing quality iron ore deposits on the Patapsco River in Anne Arundel and Baltimore Counties, Maryland.[20] Maintaining a separate Maryland estate and having to float the ore around the western shore of Maryland to Neabsco was an expensive proposition. Thus, in 1738, Tayloe went to great lengths to convince the Council of Virginia, of which he was a member, to relieve "himself and other Adventurers in Iron Mines" from port duties on iron ore imported from Maryland. The Council agreed, and ordered "the Naval Officers of the several Districts into which any Oar shall be Imported to not require the Port duties or other fees . . . for any Vessels importing Iron Oar from Maryland for the use of the Iron Works here so [long] as such Vessels do not carry any other Goods & Merchandize."[21] When Tayloe's own ore banks came up short, as they likely did in the early years when he was just getting the mines established, he would sometimes purchase ore from Stephen Onion's ore bank in Maryland. Onion was another iron-

Figure 2
Tayloe House

Williamsburg town house of John Tayloe II, who regularly resided in Williamsburg to conduct his business as county burgess and councilman. Wealthy planters were able to use their political connections and positions to influence legislation and policies in their own favor. (Photo by the author)

master who opened a large ironworks with both furnaces and forges during the 1750s or 1760s. One example when Tayloe relied on Onion occurred in 1741, when Tayloe purchased 718 tons of ore.[22] The special dispensation for importing Maryland ore would still have been advantageous under these circumstances. Highlighting the extent to which political position facilitated entrepreneurship, John Tayloe II would lobby successfully to renew this special dispensation during the 1750s.[23] This ploy surely saved the Tayloes a great deal of money, as they annually imported more than a dozen loads of ore from Maryland, averaging about 43 tons per load. In 1755, Tayloe II imported a total of 515 tons of ore; in 1756, he imported 687 tons, and in 1757, 694 tons.[24]

The works at Neabsco initially included at least one blast furnace and a water-powered grist and saw mill. The furnace sat at the base of a steep hill, while the workers' quarters sat on a relatively level ridge a short way up the Neabsco Creek. Forging, smithing, and other processing activities were conducted on the eastern side of the creek, while the more central ironmaking and domestic areas were located on the western side. The furnace measured approximately 30 to 40 foot square, and produced a competitive 600 tons per year by January of 1742.[25] At some point, probably earlier rather than later, Tayloe ordered landscaping changes that created platforms that supported the furnaces, pit mines, trench mines, and a water collection pond. Archaeological investigations in the early 1990s found that a charge ramp, a bridge house or storage facility on the charge ramp, a small forge, road traces, and an ore roasting area all also existed on the site.[26] The entire site covered an area of approximately 18 acres.[27]

The ironworks quickly became a center in Prince William County for smith's work, as many of Tayloe's neighbors regularly brought items for repair or purchased items such as tools made at Neabsco. The Neabsco smiths, such as slave "Tom a Black Smith," shoed horses, made hoes, repaired plows, and labored at a variety of similar tasks.[28] Tayloe also engaged in shipping and shipbuilding at Neabsco. He owned several "ore flats" used to transport loads of Maryland ore down the Chesapeake Bay to the Potomac River, and then up the Potomac River to the Neabsco Creek. For example, in 1741, Tayloe paid two men for "floating ore up Neabscoe," and two others for "trimming a flat" and "repairing a float."[29] Tayloe oversaw the construction of more substantial ships as well. In 1741, Whiddon Wallis worked "building a waiste to the new sloop." No records enumerate the number of ships built at Neabsco or that Tayloe owned, but when he died in 1747, Tayloe left his son a full line of "Sloops, Schooners, boats and vessels."[30] These activities became an even more important part of the industry under the management of John Tayloe II.

Figure 3
Mt. Airy Plantation

Richmond County, Virginia, seat of the Tayloe family, constructed in 1765 by John Tayloe II (1721–1779). The Tayloe family still resides in the house today. (Photo by author)

The ironworks was also a plantation that grew a variety of crops and livestock. For example, in 1740, the works produced 634 barrels of corn, 2,022 bushels of oats, 7,655 pounds of beef, 21,876 pounds of pork, 5,368 pounds of venison, and smaller quantities of wheat, peas, and beans. While most of these goods were used as provisions for the ironworks as well as Tayloe's other farms, some were sent to market or sold in smaller quantities to the works' neighbors. Tayloe sold over 175 barrels of corn, 2,000 pounds of beef, 7,000 pounds of pork, 800 bushels of oats, and small quantities of peas and beans as well.[31] Some of the 100 African-American slaves appraised as a part of Tayloe's Prince William County estate in 1747 would have been the ones conducting this agricultural work.

The Neabsco Ironworks lay almost 80 miles by land and over 150 miles by water from its owner's home in Richmond County,

Virginia. Therefore, Tayloe required a staff of managers, overseers, clerks, and laborers to run the works in his absence. It is possible that Tayloe employed some of the workers with whom he was already familiar at Bristol.[32] Either way, Tayloe employed at least eight white men year-round in 1741. These wageworkers, including Benjamin Wilkinson, John Jones, William Williams, Joseph Dean, Charles Ewell, Doctor Lawson, and John Dennis, received between £25 to £80 Virginia currency depending upon their position. Nine other white men were hired to complete particular part-time tasks including "floating ore to the landing," "work about the Funnel head," "putting in a new charge & hearth," and "building a waiste to the new sloop."

Tayloe also purchased a significant number of slaves for the works during the 1740s. Several, such as Charles, Sharper, and Pompey, had previously worked at the Bristol Ironworks and were therefore experienced, skilled, and valuable ironworkers. In all, John Tayloe I owned 100 slaves valued at more than £2800 on his lands in Prince William County, most of whom likely resided at Neabsco and assisted with some part of the ironmaking process. Many of the Neabsco slaves were likely skilled or semi-skilled, but Tayloe's use of white wage workers indicates that white men held the master craftsman and supervisory positions. The mostly adult slave population consisted of 69 men and 31 women. An additional 40 slaves worked mining, carting, and shipping iron ore for Neabsco on Tayloe's Maryland estate as well. The predominance of males in Neabsco's slave population was typical of trends in slave population in Virginia overall during the first half of the eighteenth century. While records indicate that many slaves coupled and established families (about 20% of the inventoried slaves were children), many of the men at Neabsco must have had difficulty establishing such relationships because of the small number of available partners.[33]

Travers Nash, a clerk at the ironworks between 1755 and the 1770s, recalled that the works was staffed as follows: the blacksmith shop had four slave blacksmiths, the wheelwright shop employed "a man on wages and a slave to work under him," the two ore vessels were staffed by "white skippers on wages, [and] navigated by servants and slaves belonging chiefly to the Works," "a white man on wages and a slave" drove the two horse teams, and a slave man drove the ox team. Slaves also worked cutting wood and burning coal, and as founders, fillers, ore burners, and other laborers. When John Tayloe II assumed control during the late 1740s, white skilled or semi-skilled wage workers held most of the supervisory and master craftsmen positions.[34] Later, as Tayloe's slaves learned the necessary skills, they would for the most part replace the white artisans and craftsmen. Nash went on to recall that in addition to these 50 or so workers, two additional quarters with "eight or nine hands at each with Overseers" also belonged to the works. These slaves were likely responsible for more traditional agricultural work. Since most of the rest of Tayloe's estate was so distant from Neabsco, the ironworks had to be able to produce many of its own foodstuffs in order to make the enterprise profitable. Finally, several dozen slaves resided at the "mine bank" in Maryland.[35] The slaves there mined the ore, and then floated it on "ore flats" down the Chesapeake Bay to the Potomac River, and then up the Potomac River to the Neabsco Creek.[36]

The size and gender composition of the Neabsco workforce was comparable to that of other early ironworks. In 1732, Charles Chiswell of the Fredericksville Furnace advised William Byrd II that "one hundred and twenty slaves, including women, were necessary to carry on all the business of an iron works." Chiswell himself owned about 100 slaves, although he considered himself short-handed.[37] The Baltimore Company had about 150 slaves split between its various works, while the Nottingham Ironworks employed 153 slaves. Benjamin Moore owned 54 slaves who he employed at his forge in King and Queen County, including 33 men and 21 women. The labor pattern at his forge is probably indicative of the sexual division of labor prevalent on most colonial iron plantations. The men at Moore's Forge held all the skilled occupations, such as forgeman, finer, hammerman, chafferyman, carpenter, cooper, collier, and blacksmith. The women worked in essential but ancillary services such as housekeeping, field work, and childcare.[38] Tayloe, like other contemporary ironmasters, ran his ironworks as an industrial plantation. As such, the institution of slavery was central to the business and part of what made Neabsco so profitable. Ironmaking was one endeavor where slaves made Tayloe's entrepreneurship possible. The knowledge, skills, and labor of Tayloe's slaves at his ironworks proved to be his most valuable asset and enabled the works to operate profitably until the end of the eighteenth century.

Expansion of Neabsco and Founding of the Occoquan Ironworks under John Tayloe II

John Tayloe II took over the management of the Neabsco Ironworks during the 1740s, probably upon his father's death in 1747. The younger Tayloe was 16 when the works was established in 1737, and thus he was likely already familiar with its general workings and routines when his father died. The works included a furnace, quarters for the workers and manager, and storage sheds for iron ore, coal, and other supplies and equipment. It sat on 5,000 woodland acres watered by the Neabsco Creek, which flowed into Occoquan Bay. Since Tayloe's iron ore source was located on the western shore of Maryland the works also included a full line of "Sloops, Schooners, boats and vessels" for transporting ore that had been built at Neabsco.[39] Thus, the "Neabsco Company," as his father called it in his will, was a burgeoning and intricate business when the younger Tayloe assumed control. However, the younger Tayloe recognized that the works offered other avenues for profit that his father had not explored. Soon, he added land, laborers, and other enterprises to the works that made it even more lucrative, and allowed him to engage in other entrepreneurial endeavors.

The colonial iron industry was both an integral part of Virginia's tobacco culture and also made possible by it. Iron was essential for a number of crafts, including blacksmithing, building construction, shipbuilding, and weapons production. While iron was necessary for the making of agricultural tools, iron-making was an economic activity not directly linked with agriculture. In the Chesapeake, tobacco cultivation helped facilitate the trade in iron because in many ways the two activities required similar conditions. Both required extensive human labor, provided in Virginia mostly by African and African-American slaves. Both were also generally sold on credit to British merchants in exchange for other goods, and therefore required access to regular trans-Atlantic shipping. In fact, iron and tobacco were often shipped together in a partnership that lowered shipping rates on both commodities. Tobacco was a relatively light product, even casked in large wooden barrels called hogsheads. Iron was therefore used as ballast, or a stabilizing weight that helped control the draft of a ship.[40]

The Chesapeake annually sent tons of iron to England, and in fact dominated the colonial export market in iron from the 1720s until the American Revolution. The Chesapeake's geographical circumstances coupled with the socio-economics of tobacco made the region more competitive in the colonial iron trade than northern colonies such as Massachusetts or Pennsylvania. Plentiful and accessible deep-water shipping, waterpower provided by falls, and abundant raw materials combined to make the Chesapeake the leading exporter of iron. Over a quarter of all ironworks erected in the British mainland colonies were in Maryland and Virginia, and by the 1730s, they annually exported more than 2,000 tons of pig iron each year.[41] In fact, before 1750, the Chesapeake was the only colonial region to export significant amounts of iron to the mother country. For example, in 1737, the year the elder Tayloe opened his Neabsco Furnace for operation, Maryland and Virginia produced 91% of the 2,329 tons of pig iron the colonies shipped to England (see Table 1).[42] In 1733, the 2,309 tons of exported pig iron was worth a minimum of £6,411 sterling to the Chesapeake economy, behind tobacco and grains in overall export value. By the 1760s and 1770s, the Chesapeake annually exported over 5,000 tons of iron valued at almost £30,000 sterling.[43] Tayloe's Neabsco Ironworks played a role in the Chesapeake's preeminence.

The Neabsco Ironworks was a typical colonial ironworks and used the ironmaking technology of the times. Eighteenth-century iron manufacturing was a two-step or "indirect" process. Iron ore was smelted in a blast furnace to produce rectangular chunks of pig iron about a half foot wide, four feet long, and weighing 50 or more pounds each. Pig iron was then converted to bar iron at a forge by a refining decarbonization process in which the pigs were continually heated and hammered until most of the impurities were removed. In a few cases, both processes were combined in a bloomery forge, where ore was made directly into a "bloom" of iron which was then converted to an adequate but lower-quality bar iron than produced in the two-step process. A typical furnace consisted of a stack, hearth, water-driven bellows, and casting shed. The stack was generally pyramidal in shape, 25 or more feet high and about 25 feet square at the base. The stack interior was narrow at the top and bottom and wide in the middle or "bosh." The hearth was a cylindrical reservoir into which the melted ore dropped and through which it was fed into sand molds in a casting shed.[47]

Iron manufacturing was a much more intricate process than this brief description portrays. Furnaces required huge amounts of raw materials. A typical furnace required roughly half an acre of forest to make the charcoal necessary to produce each ton of pig iron, and the average annual production of a single furnace was often more than 400 tons. Put another way, an average blast furnace required three tons of iron ore and 300 bushels of charcoal every 24 hours as long as it was operating. These materials required much preparation before they could be put to use. Iron ore had to be dug from ore banks and roasted to remove impurities. Limestone too had to be mined and stored. Acres of trees had to be felled and burned to make charcoal. Once the furnace was in blast, workers continuously loaded or "charged" alternating layers of the necessary raw materials—charcoal, ore, and limestone (or oyster shells in the Neabsco case)—through the

Table 1

Pig Iron Exports from Maryland and Virginia, 1723–1755[44]

Year	*Exports MD & VA (in tons)*[45]	*Exports All Colonies (in tons)*	*% Exported By MD & VA*
1723	15	—	—
1724	202	202	100%
1725	137	137	100%
1726	263	295	89%
1727	407	484	84%
1728	643	886	73%
1729	852	1155	74%
1730	1526	1733	88%
1731	2081	2284	91%
1732	2226	2342	95%
1733	2309	2404	96%
1734	2042	2202	93%
1735	2362	2570	92%
1736	2458	2737	90%
1737	2119	2329	91%
1738	2112	2391	88%
1739	2242	2426	92%
1740	2020	2283	88%
1741	3261	3459	94%
1742	1926	2083	92%
1743	2816	3004	94%
1744	1748	1877	93%
1745	2130	2274	94%
1746	1729	1861	93%
1747	2119	2164	98%
1748	2017	2161	93%
1749	1575	1795	88%
1750	2508	—	—
1751	2950	3229	91%
1752	2762	2928	94%
1753	2347	2740	86%
1754	2591	3274	79%
1755[46]	2132	3389	63%

Figure 4
Diagram of a Furnace

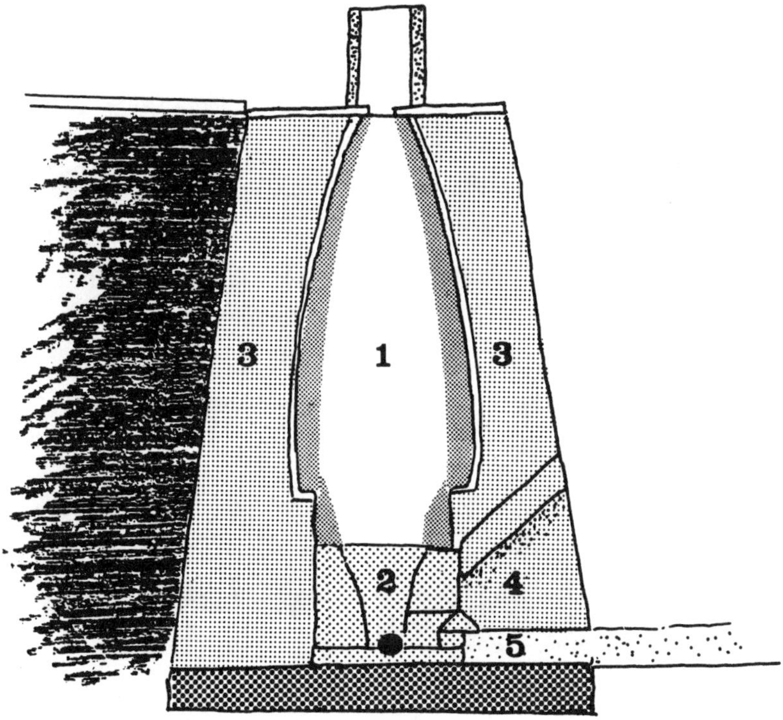

1) Upper smelting area, within the stack lining, where the flux, fuel and ore are combined; 2) Hearth, where molten iron settles; 3) The stack, usually the only part of the large furnace complex that survives; 4) The work arch; 5) The sand casting floor, where molten iron pours out into E-shaped trenches that resembled pigs suckling from a sow. (Courtesy of the *Quarterly Bulletin of the Archaeological Society of Virginia* and Ed Heite)

opening in the top. To melt the ore, the blast had to be maintained at a high temperature, most often accomplished by a water-mill-driven bellows. Melted ore dropped down into the hearth, and the fluxing agent worked to coagulate the ore's nonmetallic impurities so they could be daily drawn off as a waste product known as slag.[48]

Figures 5 and 6
Carlisle Furnace, front and rear

The Carlisle Ironworks, located in Boiling Springs, Pennsylvania, was founded in the 1760s by the Ege family. Standing on the Yellow Breeches River, this furnace stack resembles the one that stood at Neabsco. (Photos by the author)

The molten iron was poured or ladled into molds to make cast iron objects, or channeled into a series of sand molds on the casting house floor to make pigs. Both products were marketable and profitable. Cast objects might include firebacks, stove plates, tools, hollowware, or weapons. Pigs could either be marketed as is or shipped to forges or smithies to be refined into bar iron for making nails, tools, chain, or a variety of other hardware. Bar iron was also marketable, provided it met the British government's regulations regarding "the sizes of IRON proper to be imported from America for his Majesty's Service."

Figure 7
Pine Grove Iron Furnace

The Ege family opened this furnace south of Carlisle, Pennsylvania, in the 1760s and operated it for over 100 years. Similar to Neabsco and Occoquan, the furnace manufactured pig iron and cast iron products including ten plate stoves, fireplace backs, iron kettles, and military supplies. The ruins of the rest of the furnace complex can be seen to the right of the stack on the hillside. (Photo by the author)

Iron manufacturing was labor and capital intensive. The indirect process produced hundreds of tons of iron, but required night and day attention by a large crew of workmen for many months at a time. Shortages of any of the needed materials, lack of water power, or the breaking of any part of the hearth or stack immediately ended the blast.[49]

John Tayloe II understood the benefits and potential hazards of participating in the iron industry. In fact, his apparent confidence in his endeavors at Neabsco drew him to another nearby opportunity in the mid-1750s. In 1755, Tayloe and a business associate from Northumberland County named Presley Thorn-

Figure 8
Furnace Complex in Operation

This painting of a sectioned furnace shows that the furnace made up only one part of an ironmaking facility. Natural resources were charged into the top of the stack from structures above it, power was provided by water-driven mills, and sheds at the bottom provided shelter for the pig iron. (Courtesy, Pennsylvania Department of Conservation and Natural Resources)

ton contracted with John Ballendine to establish and manage a new furnace on the Occoquan River in Prince William County on their behalf. John Ballendine was an ambitious, businessman whose enterprises helped establish the small town of Occoquan, Virginia. In return, Tayloe and Thornton invested in Ballendine's Occoquan forge complex and advanced him the money to purchase 280 acres on the south side of the river. Tayloe and Thornton bought almost 1,800 acres of land for £450 Virginia currency for the new furnace site during the spring of 1755. By 1756, both the Occoquan Furnace and Occoquan Forge were in operation, as Ballendine boasted to Tayloe in May of 1756 that "Our Furnace is at Last completed greatly to my Satisfaction."

Ballendine felt that with the appropriate labor force, they could "make 1000 Tons of Piggs."[50]

The arrangement between the three men was more difficult than probably any of them had imagined. Ballendine apparently kept expanding the forge complex and became a constant drain on Tayloe and Thornton from which they saw no immediate returns. As early as November of 1756, Tayloe and Thornton advertised in the region's newspapers that "John Ballendine is no longer agent for their Furnace at Occoquan," and advised readers "not [to] trust him on their account."[51] The strained partnership apparently remained legally intact for several more years, until finally Tayloe and Thornton bought out Ballendine's interest in the whole estate. For a mere £150 Virginia money, Tayloe and Thornton acquired "the tract of land on [the] south side of Occoquan River in Prince William county . . . [with a] plantation house, the Occoquan Warehouse, and a forge containing 1,070 acres." In the interim, Tayloe and Thornton had added almost another 1,500 acres to the Occoquan Furnace. Once they cornered Ballendine into selling, they ended up with a huge furnace and forge complex with over 4,000 acres of land.[52] Between his Neabsco lands and his half of the Occoquan estate, Tayloe owned most of almost 20,000 acres of land along the Potomac River and inland between the Occoquan River and Quantico Creek in northern Virginia.

When Archdeacon Burnaby visited the Occoquan Works in October of 1759, he praised what he observed. "About two miles above Colchester there is an iron furnace, a forge, two saw mills and a bolting mill. . . . They have every convenience of wood and water that can be wished for."[53] While the complex seemed to be flourishing, disputes over its ownership and management continued. Ballendine apparently had brought in John Semple, another area ironmaster, in the spring of 1762, nine months before Tayloe and Thornton had bought him out. Thus, Semple still claimed rights in the forge complex, which sat fortuitously a quarter of a mile down river of the Occoquan Furnace. From time to time, Semple went out of his way to interrupt the work of Tayloe and Thornton's furnace, and he repeatedly turned down very lucrative offers for his interest in the works. Tayloe and Thornton discovered that even though Semple knew that they had purchased the complex from Ballendine in January of 1763, he disregarded "our possession, [and] afterwards purchased the same of Ballendine, and became possessed thereof." In desperate straits, Ballendine had complicated matters further by selling the complex to yet another party, one James Douglass, from whom Tayloe and Thornton were forced to purchase the works yet again. During the summer of 1766, Tayloe and Thornton warned the public about the status of the works and their situation: "LEST any persons inclined to purchase the lands and works on Occoquan river . . . and to prevent any pretence of ignorance of our right and claim thereto, we are induced to advise the publick that there is now of record in the Secretary's Office of this colony articles of agreement made between the said Ballendine and us, for the conveyance of the said land and works, where we were forthwith put in possession of." Anticipating further trouble from Semple, the pair cautioned *Virginia Gazette* readers, "If any will purchase of the said Semple after this notice, they will do it with their eyes open, and must stand to the consequences, as we are determined to assert our rights."[54] As late as 1771, Tayloe's manager, Thomas Lawson, could still lament, "I must think that one day or other the Occoquan Forges must fall into your hands."[55] John Semple and possibly another businessmen were still at least partly involved in Occoquan. Indeed, Semple's executors advertised the estate for sale after his death in 1788.[56]

Undoubtedly these controversies annoyed John Tayloe, a man of self-proclaimed "positiveness in opinion."[57] When Tayloe felt he was in the right, as he did in the instance of Occoquan, he could not be persuaded otherwise. It was in this spirit of persistence, self-assurance, and indignation that he began to personally manage the works at Occoquan despite the problems with Ballendine and Semple. The Occoquan Furnace, established in 1755, was located one half mile west of the town

Figure 9
Rockledge

Located in the Town of Occoquan in Prince William County, Rockledge has long been associated with the Ballendine name. Constructed ca. 1760, it is a rare example of domestic stone construction and its design is attributed to William Buckland. Rockledge has been restored several times, most recently after the destruction of its interior by fire in the early 1980s. (Courtesy, Planning Office, Prince William County)

of Occoquan on the south bank of the Occoquan River. The Occoquan Forge, established several years later, was located in the village of Occoquan on the same side of the river.[58] While the furnace sat just above the river's falls, the forge stood on "navigable waters," allowing for the easy loading and unloading of supplies, ore, and goods for market.[59]

The complex also included a grist mill, bolting mill, saw mill, bake houses, storehouses, and dwellings. The main dwelling house, where Ballendine and later ironworks managers and clerks resided, included six fashionably furnished rooms.[60] A central hall included a large walnut table that seated twelve and was set with "Chaina" bowls, tea pots, and coffee cups, as well as "Stone[ware] Plates, wine glasses, "Brass Candle Sticks," and knives and forks. Under Tayloe's and Thornton's management, at least five of his salaried workers resided in the house, three of whom shared one bedroom. One old and one new dairy, a storage cellar, a granary cellar, and a slave quarter on the "Mourmsco" Creek also existed on the property. The company also owned teams of horses, cattle, oxen, and pigs.[61]

While Presley Thornton contributed capital and likely some advice to the partnership, Tayloe was the more active partner in the daily management. Indeed, Tayloe essentially ran the two iron companies, Neabsco and Occoquan, as one operation. The two ironworks shared supplies and sometimes laborers and received their ore from the same Maryland mines. Tayloe often marketed their products without distinguishing in his accounts at which of the two furnaces, for example, the pig iron had been manufactured.[62] To further complicate matters, from one year to the next Tayloe's bookkeepers rarely closed out the ironworks' accounts, a common eighteenth-century accounting practice, and therefore it became more difficult over time to untangle the business of one works from the other.[63] Indeed, when Tayloe's and Thornton's descendants asked one Richard Parker "to make a fair judgment of the accounts of the works so that they could make a bargain," he concluded that "the accounts of the two set of works were so blended together that it appeared to me impossible to separate them."[64] Thus, any discussion of the business activities and operations of one works necessitates the inclusion of the other.

While the accounts of the two ironworks are incomplete, they do reveal the basics of the business as Tayloe ran it. First, one or both of the furnaces produced pig iron every year between 1754 and 1774.[65] Second, after the major addition of the second ironworks in 1755, Tayloe continued to add various enterprises to the two companies to make them more productive and profitable. Third, they were staffed by a combination of owned slaves, hired slaves, indentured servants, convict servants, and white overseers and clerks.

The Neabsco and Occoquan Furnaces were both productive in manufacturing pig iron for market. Less is known about the operations of the Neabsco and Occoquan Forges, as few accounts still exist. Existing receipts and accounts indicate that Tayloe productively operated at least one forge between 1756 and 1774, and sold his bar iron to both local planters and British merchants.[66] While both furnaces operated for longer periods, their main years of operation under the second John Tayloe's management that will be analyzed here are 1754–1774. Tayloe reserved some of his furnace's pig iron for the use of his forges, but exported most of it to England to be sold by English merchants with whom he regularly conducted business. Tayloe consigned his iron to at least 19 merchants during this 20 year period, but most regularly trusted London merchants James Buchanan, John Backhouse, James Russell, Robert Allason, John Stewart, and John Jordan to look out for his interests and take care of his business. Occasionally, he consigned his iron to merchants in Bristol, Liverpool, Dublin, and Glasgow as well. To protect against shipwrecks and piracy, Tayloe usually divided his annual iron production between as many as 14 ships, ensuring that at least most of his iron would arrive safely in England. Tayloe often marketed his tobacco and iron together, which was an advantageous arrangement for ships' captains as they could take on more tobacco with the iron as ballast.

Indeed, taking his two products together was a prerequisite for doing business as far as Tayloe was concerned. When he was once apparently refused in this matter, he was outraged. "[M]y Tob^{o} hitherto, so it ever shall, command fr[eigh]t for my Iron. I consider the advantage of Iron ballast to my Tobacco loaded ship . . . is greatly superior to any expense. . . . therefore upon the whole I will not ship Tob^{o} in any Ship whose Owner disputes taking Iron freight free, as usual." Tayloe directed his correspondent, London merchant William Lee, to settle his account with merchant James Russell, as he intended to "withdraw my

Correspondence for that reason only." Tayloe was a fair but demanding business associate, and refused to be "thought unreasonable, or dictatorial."[67]

The per ton prices Tayloe received for his iron fluctuated from one transaction to the next each year. Tayloe received the highest average prices per ton for his pig iron during the mid-1750s and late-1760s, while the early-1760s and 1770s proved less lucrative. These fluctuations seem to coincide with depressions in Virginia's tobacco economy during these years.[68] One shipment of Tayloe's iron commanded an all-time high of £7 12s. 10d. per ton in 1759, while others brought a disappointing £3 to £4 per ton during the period 1758–1760. On average, Tayloe usually received between £5 and £6 sterling per ton for his pig iron (see Table 2). This iron price pattern roughly coincided with fluctuations in the tobacco prices Tayloe received, which peaked in 1757 and 1769. During depressed years then, iron acted as an important hedge against the uncertainties and difficulties of the tobacco economy. While iron and tobacco prices were generally simultaneously depressed, the combined income Tayloe received in depressed years was far more than he would have received for tobacco alone.[69] For example, when Tayloe's tobacco brought only £4 16s. 10d. per hogshead for 153 hogsheads he sold in 1763, he received £6 14s. 6d. per ton for 214 tons of pig iron. His total income from both commodities that year reached just over £1957 sterling. Had Tayloe been dependent on tobacco alone, he would have made just £712.

The prices Tayloe received for his pig iron between 1754 and 1774 were generally in line with those at other ironworks. Between 1728 and approximately 1760, Charles Chiswell and Alexander Spotswood operated the Fredericksville Furnace in Spotsylvania County, Virginia. During the 1730s, Chiswell advised William Byrd II on the conditions necessary to establish a successful ironworks. If Byrd followed his advice, Chiswell assured, he could make 800 tons of pig iron per year and sell it for £6 sterling per ton.[71] Bigger operations, like the Baltimore Company of Maryland, were able to command even more for

Table 2
John Tayloe II Tobacco and Iron Exportation, 1751–1774[70]

Year	*Hogsheads of Tobacco*	*£ Sterling Received*	*Tons of Pig Iron*	*£ Sterling Received*
1751	45	£27	—	—
1752	53	£303	—	—
1753	59	£246	—	—
1754	97	£430	109	£657
1755	101.5	£781	40.5	£256
1756	75.5	£626	416.5	£2549
1757	50†	£478	124	£706
1758	194†	£865	281†	£1049
1759	89†	£654	450†	£2343
1760	125	£986	258†	£1112
1761	127†	£932	110.5	£566
1762	129	£635[a]	289	£1617
1763	153†	£712	214†	£1245
1764	119	£696	60	£338
1765	60	£517	24†	£97
1766	75	£540	129†	£667
1767	42	£320	67†	£603[a]
1768	36	£292	35†	£179[a]
1769	64†	£606	93†	£550
1770	23	£204	54	£325
1771	28	£198	0	0
1772	47†	£115	120†	£365
1773	19	£94	76†	£167
1774	12†	0	40†	£20

† = portion of annual amount was lost at sea
[a] = no value recorded for portion of exports, or value listed without tonnage figures

their iron. Established in the 1730s by five prominent planter-businessmen—Daniel Dulany, Dr. Charles Carroll, Benjamin Tasker, Charles Carroll of Annapolis, and Daniel Carroll—the Baltimore Company became one of the largest and most successful ironworks in eighteenth-century America. Initially, the Carrolls figured that an annual production of 500 tons bringing

£8 sterling per ton would allow them to begin seeing a profit in just two years. However, £6 to £7 sterling per ton proved to more realistic.[72]

Comparison with other furnaces is instructive regarding production levels as well. The amount of pig iron Tayloe's ironworks jointly manufactured each year fluctuated from a high of 450 tons in 1759 to just 35 tons in 1768. Furnace repairs, inclement weather, and insufficient raw materials could and occasionally did completely ruin Tayloe's chances of achieving a productive blast. The Neabsco Furnace produced an average of just under 300 tons of pig iron per year during the period 1756 to 1763—the furnace's most productive period for which there are records. While records for the Occoquan Furnace are less complete, it is probably safe to assume that Tayloe operated it at about the same levels. Thus, Neabsco and Occoquan, which Tayloe operated as one company, produced about 500 tons of pig iron per year from their two (sometimes three) furnaces. Meanwhile, the Baltimore Company produced 376 tons in 1750 from one of its furnaces.[73] The Principio Company produced 820 tons of pig iron in 1736 from its two furnaces.[74] Thomas Jefferson provided much information on Virginia's iron industry in his famous 1787 *Notes on the State of Virginia.* Jefferson recalled that David Ross's Campbell County Oxford Furnace manufactured an astounding 1,600 tons of pig iron per year. John Ballendine made 1,000 tons per year at his Bear Garden Furnace in Buckingham County; J. Calloway, John Miller, and Isaac Zane each produced about 600 tons of pig iron annually.[75] The ironmasters Jefferson described benefited from the plentiful natural resources located in western Virginia, where the iron industry shifted after the Revolution. Thus, while Tayloe's production levels might have been on the low end of this comparative sample, they were still in the average annual range of about 500 tons of pig iron.

To determine whether or not Neabsco and Occoquan were profitable, all marketing and production costs must be deducted from Tayloe's gross annual receipts. For example, in 1756, Tayloe produced 416.5 tons of pig iron at Neabsco. He consigned it to two London merchants in eleven different ships. In all, he received £2549 5s. 1d. for his pig iron, or approximately £6 12s. per ton. Tayloe's incomplete records make it difficult to assess his production and marketing costs, but they were likely comparable to Charles Carroll's 1751 Baltimore Company expenses. In that year, Carroll sold 376 tons of pig iron to London and Bristol merchants, and received £2324 10s. 2d., or about £6 3s. per ton. It cost £3 2s. 3d. to make a ton of Baltimore pig iron, so Carroll's production costs on 376 tons were £1170 6s. Including freight, wharfage, insurance, commission, and other shipping charges, it cost Carroll about £545 to sell his pig iron. Thus, it cost Carroll £1715 to make £2324, leaving him a profit of over £600.[76] Carroll's shipping expenses that year included extra charges related to an unusual set of circumstances. The ship carrying his iron to Bristol, the *Charming Molly,* ran ashore in Wales and her cargo was thereafter sold for salvage. Carroll then paid additional charges to reship his freight, creating total extra expenses in 1751 of more than £127.[77] This meant that Carroll's expenses in a non-catastrophe year were just under £1600. Assuming Tayloe's shipping and production expenses were relatively similar, he would have invested around £1600 in 1756 to make £2549, leaving him with a hefty profit of nearly £1000 from his ironworks alone.

The profitability of the ironworks allowed Tayloe to expand his ironmaking activities and diversify into new enterprises. After acquiring the entirety of the Occoquan Company in 1763, Tayloe set about expanding the ironmaking capabilities of the estate. He purchased 500 acres between the Occoquan River and Neabsco Creek to connect the two ironworks together in 1765. Along with his partner Presley Thornton, he added over 2,500 acres to the Occoquan Company during the late 1760s.[78] These lands provided ready access to several essential raw materials, especially timber and surface iron ore deposits. Tayloe also added to the labor forces at the two sites. As of 1750, Neabsco had approximately 100 workers, mostly slaves and a few indentured servants, who did everything from skilled ironmaking jobs to agricultural field work. This number remained fairly constant throughout Tayloe's lifetime, as he only occasionally purchased slaves for Neabsco, such as a new slave skipper in 1771. As a new endeavor, however, Occoquan was a different story. Tayloe helped build Occoquan's labor force from scratch. In 1755 and 1756, he purchased Tom, Jack, and 14 other slaves whose names were not mentioned. Between 1758 and 1760, he added another 23 slaves, and also began hiring from area slaveholders between three and five other slaves for various seasonal work each year. When the estate was inventoried after Presley Thornton's death in 1771, 69 slave men, women, and children valued at £3185 lived and worked at Occoquan.[79] Together, the Neabsco and Occoquan companies included over 170 slave laborers and almost 20,000 acres of prime tidewater lands on the eve of the American Revolution.

During the early 1760s, Tayloe began working on other new enterprises as well. Tayloe built "a [new] merchants mill" at Neabsco, which operated in addition to a grist mill and saw mill already on the property.[80] Water mills encompassed some of the most advanced technology of the eighteenth century, and required technical skills and entrepreneurial drive to establish and operate. Operating various types of mills at once allowed Tayloe to extend their use and value beyond the limited season for grinding grain. For example, as long as water was available to power the mill, timber could be sawed into plank for market or staves for flour barrels.[81] Another saw mill existed at Occoquan, from which Tayloe regularly sold various kinds of plank from the 1750s through the 1770s. Indeed, Ballendine exclaimed to Tayloe that he "hope[d] you won't ingage any more plank to be sent by us as the demand here is so great [that I] can venture to tell you I cou'd sell from our own Landing four times as much as our mill can saw."[82] By the early 1770s, however, the "timber was scarce and at some distance from the [Occoquan] Mill," and Tayloe apparently abandoned it.[83] Also, by 1771, the older mills at Neabsco began to fall into disrepair, so he constructed new grist mill. The new mill ground Tayloe's own corn and wheat,

Figure 10
Ruins of Menokin

Using wealth generated from his combined tobacco and iron enterprises, John Tayloe II constructed this mansion in the early 1770s a short distance from Mt. Airy as a wedding present for his daughter Rebecca and her new husband Francis Lightfoot Lee, signer of the Declaration of Independence. The Lees died without children in 1797 and the property passed through multiple owners in the nineteenth and twentieth centuries as it declined. In 1995, the Menokin Foundation acquired the house and is currently working to restore and learn from it. The roof was added several years later to protect the ruins from the elements while restoration is underway. (Photo by the author)

his neighbor's "Country corn," and wheat for the market at Alexandria.[84] Tayloe apparently did quite a business with his mills, as his manager Thomas Lawson estimated that they made "not less than between £7 & £800" in profits during 1774 and 1775. Lawson mused that milling was "a very pretty business could it be carried on to that perfection I would chuse to have it."[85]

Tayloe initiated some other new enterprises during the 1760s as well. Travers Nash recalled that Tayloe established "a little

Figure 11
Tayloe Mill

The Tayloe family operated numerous mills on all of their lands, including those in Prince William County. This mill, built in the eighteenth century, still stands on lands near Mt. Airy in Richmond County and operated into the twentieth century. Milling was a critical part of the Tayloes' business activities in Prince William County, especially after the closing of the ironworks at the turn of the nineteenth century. (Photo by the author)

Figure 12
Tayloe Mill, interior

Interior, showing the mill stones that would have ground the grain. (Photo by the author)

retail store of Wet and dry goods" at Neabsco. While John Tayloe I had been heavily involved in merchant activities, the younger Tayloe's merchant role was generally more limited. However, circumstances at Neabsco made it advantageous for Tayloe to open his retail store. Instead of paying his white wage workers a salary, after 1761, Tayloe "principally pa[id] off Wagemen . . . [in] provisions."[86] During the 1760s, Tayloe regularly received goods and provisions intended for sale to his workers. While this was not likely a high-volume business, Tayloe must have made some money from the higher prices he could charge at the store his workers had no choice but to patronize.

At about the same time, Tayloe began building new ships at both Neabsco and Occoquan that proved to be valuable assets to all of his business interests. He used ships to transport goods to market and between his numerous plantations. In 1761 he built two ships at Neabsco. Thereafter, Occoquan became the center of Tayloe's shipbuilding activities. During the 1760s and 1770s, Occoquan ship carpenters built an unknown but likely moderate number of schooners, sloops, and other vessels. For example, in 1766, Tayloe sold a small ferry boat to William Brockenbrough from the Occoquan Company. In 1771, Lawson examined a recently received shipment of supplies, and was relieved to notify Tayloe that their awaited "Sail Duck & Rigging for the vessels engross more than two thirds of the amount."[87] Lawson later complained to Tayloe that "the new Sloop . . . has been ready for launching this three months almost," but animosities between skippers Lawson hired delayed the launching. As late as the 1780s, when ironmaking at Occoquan had been largely abandoned, slaves were still engaged in shipbuilding there, as a half dozen slaves were working on "A Schooner on the stocks."[88]

Tayloe's ships regularly visited Hobb's Hole (now Tappahannock), Baltimore, Alexandria, and Norfolk. During the 1760s and 1770s, at least eight sloops and schooners named *Betty, Jenny, Occoquan, Polly, Beckey, Nancy, Elizabeth,* and *Auogon* serviced Tayloe's various estates.[89] Tayloe's shipbuilding and shipping activities nearly made him a one-man "shipline."[90] Furthermore, his ironmaking activities contributed to his ability to successfully participate in the shipbuilding industry, as each vessel required tons of iron and timber. Indeed, Chesapeake shipbuilders nearly dominated colonial shipbuilding in the same manner they did colonial ironmaking. Numerous navigable rivers, abundant woods close to waterways, and ample swamps to supply the industry with timber, tar, pitch, and turpentine meant that the Chesapeake in the years before the Revolution annually produced about 50 ships. Only Massachusetts outdistanced the Chesapeake in pre-Revolutionary shipbuilding.[91]

From the 1750s until Tayloe's death in 1779, ships played a central role in allowing Tayloe to communicate with his distant overseers and coordinate the activities of his various enterprises. Tayloe sent pig iron to markets at Rappahannock and Norfolk in his sloops and schooners, and Ballendine delivered pine plank from Neabsco to one Mr. Pettite via "our Little sloop."[92] The schooner *Occoquan* traveled down to Presley Thornton's Northumberland estate to retrieve corn for the ironworks' slaves, while the *Nancy* sailed to Freestone Point at the confluence of the Neabsco Creek and Occoquan Bay "after shells" to use as the fluxing agent for the next blast.[93] The *Betty, Beckey,* and *Elizabeth* frequently carried goods between the Maryland plantations and the ironworks, and all the vessels assisted in moving slaves, correspondence, and provisions between the diffuse corners of an estate that included over a dozen plantations in nearly as many Virginia and Maryland counties.[94]

Renting lands comprised a final entrepreneurial activity connected to the ironworks in which Tayloe engaged. Travers Nash remembered that during the 1760s, the ironworks' managers picked up the additional responsibility of "taking the collection of rents and granting leases" to some of Tayloe's numerous tenants.[95] For example, in 1761, Tayloe and Thornton added a small parcel of land to their ore banks in Maryland. While these lands provided essential raw materials for the ironworks in the form of iron ore, they were apparently "Hilly & poor Land destitute of Clapboard Timber." A small portion of the lands were useful for cultivation, and a dwelling house, tobacco house, apple orchard, and slave quarter existed on the property.[96] By 1770, Tayloe realized a way to derive even more profit from these lands. Richard Croxall, a manager for the Baltimore Company and one of Tayloe's business associates, recommended a man of "honesty and the greatest industry" to whom Tayloe could rent "that plantation where the mine bank is." Thomas Lawson met with the man and seemed to agree with Croxall's recommendation. Lawson arranged to rent the plantation for nine years for a total of £315, reserving the iron ore for Tayloe, and felt, "In the doing of which I'm confident that the place will be of more than double the value." Lawson followed suit with other tracts of Tayloe's Maryland land as well.[97] Making such arrangements was a typical part of the business for Lawson by the 1770s, as he also took over responsibility for managing Tayloe's Kittockton (or Ketoctan) Manor tract containing 4,000 acres in Loudoun County.

None of Tayloe's entrepreneurial activities would have been possible or successful without the oversight and labor of his numerous managers, clerks, indentured servants, and slaves. On an estate as extensive as Tayloe's, trustworthy and diligent overseers, managers, and agents were essential. Tayloe's most important agent was Thomas Lawson, manager of the Neabsco and Occoquan Ironworks. Lawson came to Neabsco in 1755 and worked there until his death in 1785. Lawson was more than an overseer—he conducted business for Tayloe on an estate-wide level. Throughout his employment, Lawson received £100 currency annual salary for his management of both iron companies. Lawson had many responsibilities. Besides overseeing the production of iron by the works, Lawson also handled the Kittockton leases and rents, coordinated supplies and slave labor for and between the ironworks and Maryland estates, supervised shipbuilding activities and mill operations, negotiated labor contracts and other personnel issues with other white employees, and corresponded with Tayloe's numerous business associates on a variety of matters. Lawson continually worried over the immense debt owed to Tayloe's enterprises, and as often as Tayloe wondered what could be achieved "could I get half that is due to this place and Occoquan."[98] On his frequent travels, Lawson always got the latest market information for Tayloe and negotiated deals with other businessmen when possible. On one trip to Baltimore, Lawson arranged with a Captain Grieg to take "30 tons your Iron, freight free, and to wait for it till the 15th of January." Lawson excitedly reported to Tayloe that the Maryland furnaces were not able to meet the demands of the waiting ships, and was "not sorry . . . as ours will be in greater demand."[99] Lawson realized his stake in Tayloe's success, and regularly assured Tayloe that he would put forward his "best endeavours for the best."[100]

Lawson hired a number of employees to assist him in running the works. Travers Nash served as the head clerk of the works, and felt he "did under Lawson's direction transact the chief part of the business of the works." Nash also arrived at Neabsco in 1755 and worked there until 1775, when he and the Occoquan overseer, Posey, got into "a fisty cuff battle . . . wherein the latter came off victorious, and entirely drove the other off the field." Lawson described Nash as a difficult person with a "most wicked and revengefull disposition," and his injuries from the fight apparently prevented him from working again.[101] While head clerk, Nash received between £30 and £50 currency each year. Nash recalled that a number of other assistants worked at Neabsco as well, some of whom he felt were not entirely necessary. These men, such as Posey, worked in a more typical overseer capacity and received £20 per year for their services.[102] Thomas Lawson Allison, William Lawson, and William Holburne each served in turn as head clerk at Neabsco after Nash became incapacitated.[103]

A number of other white workers served the ironworks as well. Lawson hired several white skippers, for example. As early as 1755, "white skippers on wages" operated the ore flats. In 1771, Lawson found himself trying to smooth over the personalities and egos of several of his captains. Captain Gallahue was "determined to be master of the new Sloop," while Captain Steele "of course expects the *Beckey* as his lot." Captain Dick piloted the *Polly* and Captain Grant navigated the schooner *Betty*.[104] Not all the ships were piloted by white skippers. As of 1778, William Lawson recalled that two schooners were navigated by slave crews while one sloop sailed under a hired skipper and slave crew. Increasingly, Thomas Lawson found that there "[was] no such thing as getting White Skippers [anymore], and if I could they are generally as bad at the Blacks."[105] As with the many other enterprises at Neabsco, Tayloe's slaves began to replace white workers and servants as they learned the appropriate skills.

Tayloe regularly brought convict and indentured servants to the works to serve their terms. When Occoquan opened, Tayloe immediately purchased £90 worth of servants to assist in the production of iron. Convict servants served terms of up to 14 years, while indentured servants served an average of seven years. Some 50,000 convict servants were brought to America between 1718 and 1775. Chesapeake planters received a majority of these convicts, most of whom originated from London jails.[106] Tayloe purchased convicts individually throughout his lifetime, many of whom ended up working at his ironworks. At least seven of these men attempted to escape from service at Neabsco, and apparently repeatedly had the opportunity to do so while working as skippers or crewmen on Tayloe's ships. Timothy Carpenter had only been in Virginia for nine months when he ran away from the schooner *Billy*. Seventeen year old Arundale Carnes ran away from Neabsco in a canoe in August of 1766, and was likely to try to pass for a doctor, a gentleman's body servant, or a sailor. Tayloe also owned several indentured servants, such as the "Ironman" William Gill and the forge carpenter John Milborn, both of whom also ran away from the works.[107] Joseph Fosset and Joseph Stephens both worked at Occoquan as wagon drivers. Fosset (or Fositt) ran away from Occoquan in 1765 with three slave carpenters.[108]

Proportionately, Tayloe's slaves absconded less frequently than did his convict servants, although the majority of Tayloe's slave runaways fled from the ironworks rather than from other parts of his estate. Skilled slaves, such as ironworkers, were more likely to run away than unskilled slaves because their daily contact with whites gave them the skills, knowledge, and confidence to operate successfully and comfortably in the white world.[109] For example, Frank and George ran away from Tayloe's Maryland mine banks in 1754, and miner Dick ran away with a servant man named Richard Wiggan in 1759.[110] Pompey ran away from Neabsco in 1757 and made it across the Potomac to Charles County, Maryland, before he was captured. Another male slave belonging to Neabsco got ten miles from the works before he was "taken up" by a local planter.[111] Likewise, Anthony got ten miles from Occoquan before being caught by James Calk in 1757, and Occoquan carpenters Billy, Billy, and Sampson escaped with a convict servant named Joseph Fositt in 1765.[112]

The Occoquan carpenters and Joseph Fositt were all recaptured at some point after their escape, but ship carpenter Billy repeatedly attempted to obtain his freedom. In August of 1768, he again ran away from Occoquan. He was described as "a very well set fellow, about 5 feet 8 or 9 inches high, chews tobacco, can play on a violin . . . and is by trade a ship carpenter." Billy took with him "a black cotton velvet jacket, two white shirts, one pair of brown and a pair of check linen trousers, an old fine hat bound round the brim with black tape, two pair of shoes, and plated buckles." During his previous escape attempt, he ran to Carolina, "where, by virtue of a forged pass that some good natured person had wrote for him, he had travelled without much interruption." Lawson supposed that he was heading for South Carolina, where he expected to be free and work as a ship carpenter.[113] Billy's plans were apparently thwarted in 1768, as by February of 1769, Lawson again advertised that Billy ran away. In October of 1768, "Billie" ran off with Scipio, a hired slave belonging to John McMillan. "They crossed Patowmack river together, in a schooner's boat, to the Maryland shore, where they left her, and have, from that time, kept themselves undiscovered." Lawson supposed that since Billy had just been brought back from Carolina, where "he had travelled as a freeman[,] it is more than probable that if he is not now engaged by some ship builders to the northward, that he will endeavour to get on board some craft, bound for Charles-Town, or to some place in Carolina, where he expects to be free."[114] Billy's skills undoubtedly helped him find employment and pass as free. Lawson described him as a masterful ship carpenter, who could "not only . . . repair, but . . . [also] build all sorts of small craft."[115]

Ironmaking was the nonagricultural pursuit that made the most use of slave labor, and Tayloe's works were unquestionably dependent on slaves. Approximately 170 slaves worked and lived at Neabsco and Occoquan. Tayloe mostly used his own slaves at the ironworks, but did regularly hire a small number of others. For example, between 1759 and 1779, Tayloe annually hired three to five slaves from Thomas Lawson for £12 10s. each, including "Tippin & his wife," Nan, and Paisley. In 1759, Presley Thornton paid just over £39 "for the hire of 2 negro carpenters for 9 months."[116] The majority of Tayloe's slaves at Neabsco and Occoquan, such as Billy, were skilled. In fact, even though white men sometimes held the head positions, slaves possessed the full spectrum of skills related to the works. William Lawson, Neabsco bookkeeper between 1778 and 1783, recalled that the following labor arrangements were in place when he arrived at the works:

> [F]ourteen Wood cutters besides white men hired at different times, Six colliers and a hire[d] Overseer—five blacksmiths[,] two ship Carpenters, two Wheelwrights, two Coopers, two shoemakers, one tanner[,] three house carpenters, a grist Mill kept by a Negro, a Merchant Mill kept by a hired Miller and a Negro, two horse teams and One oxteam drove by negroes, six or eight hands employed in Manufacturing Cloth and linnen . . . a Schooner navigated by five hands, a Sloop Navigated by a hired Skipper and four hands, a schooner Navigated by four hands, two hands employed in a flat bringing Ore to the landing. . . . I cannot

> exactly say how many hands were employed at the Mine bank in diging [sic] ore, hauling it to the landing and working the plantation there, tho' I believe there could not be fewer than twenty.

Skilled workers were essential to the various enterprises at Neabsco and Occoquan, and each furnace depended on the whole pool of skilled laborers available at both works. For example, Lawson indicated that "whenever either furnace was going into blast it was assisted by hands from the other in preparing the Stock."[117] Lawson's description also illustrates that both men and women performed vital and gender-determined functions at the iron plantations. As early as 1747, 31 women worked at Neabsco and at least nine women labored at the Maryland ore banks. In addition, the Occoquan labor force included 25 females in 1771. Slave men worked as founders, furnace keepers, fillers, blacksmiths, millers, colliers, miners, woodcutters, carpenters, and skippers, while slave women worked as spinners, weavers, cooks, laundresses, dairy maids, and domestics.[118]

These skilled slaves increasingly worked without direct supervision, spent considerable time by themselves or with other slaves, and had at least some ability to move about the region on business. William Lawson recalled that two schooners and one of the ore flats were operated by slaves without white skippers. When Presley Thornton died, the appraisal of the estate was delayed because "four of the negroes which navigate the Sloop were not then arrived." In 1771, a number of slaves made what was likely a regular trip from Occoquan to Maryland and back alone.[119] Around Christmas that same year, "Potomack Dick" went to the town of Dumfries alone on company business. He "got so very Drunk . . . the day after the last deep snow fell, that he could not reach home, but perished on the road side in the night time, there being no one nearby to assist him."[120] While Dick's freedom resulted in tragedy, it is representative of the degree of autonomy and self-determination Tayloe's skilled slaves received in return for their talented work. Furthermore, such high degrees of mobility among the skilled slave population may partly explain the relative infrequency with which Tayloe's slaves ran away.[121]

The slaves at the ironworks married and raised children, although men dominated the overall population. Sixty-nine slaves resided at Occoquan, including 38 men, 10 women, and 21 children. Judging by their appraised values, probably 8 of the women were of childbearing age. By the Revolution, the Neabsco slave population of more than 100 slaves probably had a more gender-balanced ratio than the two to one sex ratio of 1747. While most of these slaves were probably country-born, a number had either been born in Africa or named by African parents, including, Collier Quamina, Congoe, Quagua, Juba, Old Quagua, and Cuffy. Two of these men, Quagua and Cuffy, were the most valuable slaves at Occoquan, each appraised at £120.[122] While it is speculative, it seems that Occoquan slaves regularly named their children after fathers or other immediate family members. For example, Old Daniel, Nanjemoy (Maryland) Daniel, and a child named Daniel all lived at Occoquan. Likewise, Great Ben, Schooner Ben, and a small child named Ben resided at Occoquan. Old Quagua and Quagua were likely related. Founder James was listed next to a younger James, who perhaps was a relative learning the founder's skills. A similar pattern occurred at Neabsco as well. While direct family relationships are not discernible, the repeated use of the names Robin, Boatswain, Billy, Jack, Bristol, and Ralph among the Neabsco slave men probably indicates some degree of kinship ties. In analyzing the naming practices of slaves in Mississippi, South Carolina, and Georgia, Herbert Gutman found that slaves frequently named children after fathers in a fashion that affirmed the important cultural role of slave fathers in opposition to their owners' misunderstanding of slave beliefs and practices.[123] All of the Neabsco and Occoquan slaves belonged to an absentee planter, worked largely unsupervised, and spent most of their time with other African Americans. It is therefore probable that the slave communities of Neabsco and Occoquan maintained and practiced a number of beliefs and traditions, some of which were likely African-influenced, that differed from those of their owners.[124]

Challenges of the Revolutionary Era

Even with Lawson's constant supervision of the business and his slaves' and servants' usually steady labors, Tayloe experienced a number of difficulties that illustrate the risks of operating an enterprise as complex and intricate as an ironworks. Problems with runaways, delays in receiving supplies, malfunctioning equipment, and uncooperative weather regularly interrupted the smooth functioning of Tayloe's iron companies. The 1770s was a particularly troublesome decade for Tayloe's iron enterprises, as Neabsco and Occoquan regularly experienced setbacks that resulted in some of their most unproductive blasts. During the years for which records exist (1770–1774, 1777–1778), the two ironworks together produced a meager 583 tons of pig iron worth £878 3s.[125] The works manufactured no pig iron at all in 1771.

By 1770, Neabsco had operated for over 40 years while Occoquan had been in blast for 15. The works shared the same lands and thus timber and ore supplies, which were being depleted from years of lumbering and mining activities. Throughout the early 1770s, Thomas Lawson repeatedly experienced shortages of coal and ore. He informed Tayloe that Neabsco could have been in blast during the early winter of 1771 had that works not had to supply "Occoquan with a good deal of coal last spring." Afterwards, Occoquan was still short of the coal it would take to put that furnace into blast.[126] Lawson fell short of the necessary amount of iron ore in 1771 and 1775. He sent a sloop to Baltimore in 1771 to fetch a load of ore, but the vessel "vexatious[ly]" returned empty. In 1775, Lawson feared "not having a sufficiency of Ore to carry her [Neabsco furnace] on the course of the Winter," and intended to send to Baltimore to see what might be available there. Years of mining on the Anne Arundel and Baltimore County lands drained the ore deposits there. Lawson admitted "that the Bank is more scanty than usual, but I cannot help thinking there has been some misconduct also."[127] Lawson acknowledged the impact of this resource depletion on milling activities as well. He told Tayloe, "The logs in the woods are now at such a distance from our saw mill, and the teames being otherwise necessarily engaged has hitherto prevented the plank being all of it got ready and sent to Mr. Thornton as I intended." By the end of the decade, someone recalled that the Occoquan saw mill was abandoned because "[t]he timber was scarce and at some distance from the Mill," and was therefore no longer profitable.[128]

Deteriorating facilities at the ironworks further interrupted production. In 1771, Tayloe installed new hearths in the furnaces at both Neabsco and Occoquan. In October, Lawson reported to Tayloe that the new hearth had just been completed, although "the Stones that compose it were a good deal too small, yet am in hopes will endure a pretty long blast." Besides the hearth, Neabsco's "furnace Wheel and shaft . . . were as much out of order as could be, & indeed the Shaft was very rotten, as are also the bellows."[129] In all, the repairs to Neabsco's furnace took several months to complete. At just about the same time, the Occoquan Furnace required a new hearth and bellows, which a white craftsman named James Rowlings repaired. Thus, in 1771, neither furnace produced pig iron. Unfortunately, Lawson's hopes for the Neabsco hearth stones went unanswered; by September of 1775, he was again overseeing the "putting in [of] a New Hearth at Neabsco."[130] Tayloe also built a new merchant mill in 1771 to replace the older ones built during the 1750s and 1760s. Lawson worried that the mill was taking too long to complete, and complained that "Chapman the Millwright has had so many other Jobs on hand . . . [that have] prevented him giving the attention to the building of ours, that I could have wished." Lawson was pleased with the quality of the mill's construction, but surprised at the inconveniences of building a new facility. "I could not have thought that the building of such a mill would have taken up so much time, nor have employed so many hands as it has already done, the difficulty in getting stones & the quantity the walls have taken & of lime also would surprise any one."[131]

Severe weather wrecked further havoc on the ironworks. In the spring of 1772, the Neabsco Creek flooded after a thunderstorm and carried away the main furnace and mill dams. Tayloe apologized to William Lee, a merchant associate in London, for the small amounts of iron he had sent. "[I] am sorry Iron sell no better tho at present I have non[e] owe[in]g to looseing my Neabsco dam, which is now repairing."[132] The fall of 1775 brought a series of thunderstorms to the region that again flooded the creek. Lawson had replaced the dams destroyed in 1772 with tumbling dams and flood gates that could more easily withstand the force of flood waters. He felt "lucky, very lucky" that the works "did not undergo the fate of March 1772." Unfortunately, "The Occoquan furnace and saw Mill dams both gave way at their weakest places," although they were quickly repaired.[133] Unexpectedly cold weather could have an adverse effect on shipping the iron. In 1774, Tayloe informed Lee that his latest shipment of iron was "unluckyly prevented by frost from getting [sic] Iron round in time for Brown."[134]

The uncertainties of trans-Atlantic shipping also plagued Tayloe's enterprises throughout his lifetime. Between 1757 and 1774, Tayloe lost at least 369 tons of pig iron and 182 hogheads of tobacco while they were in transit to England.[135] Tayloe's bookkeepers frequently noted "lost" or "taken" shipments, although they did not usually explain the circumstances of each ship's misfortune. Likely, some of Tayloe's shipments fell victim to international rivalries while others were lost to the sea. One occasion when this nearly happened occurred in June of 1772. Tayloe had placed 25 tons of pig iron on board Captain Fox's vessel, *Matty*. At Norfolk, Fox's ship "sprung a leak," delaying Tayloe's shipment to William Lee.[136] Fortunately, Tayloe usually divided his annual shipments of both iron and tobacco between numerous vessels. However, in some years, such as 1773 when all 76 tons of his pig iron were lost, such precautions failed to protect Tayloe's interests.

The American Revolution provided a whole array of new opportunities and challenges to Tayloe and his ironworks. As early as September of 1775, Tayloe began supplying James Hunter's Fredericksburg Rappahannock Forge "with a sufficiency of Pig Iron."[137] Hunter was one of several ironmakers who focused their forging activities on the manufacturing of arms and munitions before and during the Revolution. Tayloe himself produced military equipment on behalf of the American cause, and was paid £705.8.3 for "sundry cannon Ball, plank, and Pigg Iron furnished for the Navy in Portsmouth" in June of 1776. He also supplied the Prince William County militia with almost £21 of "sundry tools" in 1778.[138] These are a few of the many instances of Tayloe's participation in the revolutionary cause before his untimely death in 1779. War production revived business at Tayloe's works during the late-1770s and early-1780s. At a minimum, the Prince William County works produced 104 tons in 1777, 189 in 1778, 313 in 1780, and 269 tons in 1782. In 1782 alone, Tayloe received over £1000 for his pig iron.[139]

The war brought Tayloe new business opportunities, but also increased the risks associated with operating an ironworks. As early as 1775, rumors circulated regarding sightings of British vessels in the Chesapeake Bay, and Lawson admitted to Tayloe, "I am every day in dread of hearing that Cutters are in the Bay and Rivers and will seize our vessels. If this should ever be the case, farewell all Virginia Iron Works."[140] Lawson's fears materialized. Tayloe nearly lost one of his schooners in 1777, but Captain Grant of the *Betty* "very luckily escaped from a M[an] of war, which chased and fired two shots at him as he came down the Bay." William Lawson recalled that by the 1780s, "the Navigation of the Bay and Rivers afterwards came to be shut up" because so many vessels had been taken or lost.[141] Tayloe experienced such loss in 1781, when the British seized "a vessel loaded with 100 barrels of corn . . . from Nanjamy [Nanjemoy, Maryland]" on Port Tobacco Creek.[142] Such conditions made it difficult to import the necessary iron ore from the Maryland ore banks, as the "hazard & risque attending the same during the war in all probability [kept] the owners & managers from attempting it."[143] Tayloe managed by occasionally risking those trips and by using the lower-quality ore deposits locally available on his Neabsco lands.

Tayloe experienced a final familiar problem during the Revolution. Commissioners examining various estate settlement issues for Tayloe's executors explained the situation: "It must be well remembered that during this time [late-1770s] our Rivers & bays were often infested by the British Vessels and that our Slaves were escaping to them by every Opportunity."[144] The British began blockading the Chesapeake Bay in 1777, and increasing numbers of slaves fled to British ships and patrols. The British welcomed the runaways and even put them to work in various capacities. British fleets returned to harass the Chesapeake during the spring of 1778, summer of 1779, and fall of 1780, but the region sustained the worst damage in 1781 when the British Army "raged in Virginia" for over six months to the tune of approximately three million pounds sterling.[145] It was during 1781 that a mulatto slave named Bill ("alias 'Will'; alias 'William'") belonging to Tayloe's ironworks was indicted for "'Treason,' in aiding and abetting, and felloniously and traiterously waging & levying war against the Commonwealth, in conjunction with divers enemies of the same, in an armed vessell." Bill apparently convinced the court that he had not volunteered to join enemy forces. Instead, he claimed "he was taken in an oyster boat and forced agst. his will, and that he never took up arms agst. the Country, and no Positive Proof that he certainly did aid or assist the Enemy, of his own free will." The court condemned Bill to death by hanging, but Governor Jefferson pardoned him shortly before his scheduled execution. Bill's story and Judge Henry Lee's argument that "a Slave . . . cannot commit Treason agst. the State, not being admitted to the Priviledges of a Citizen, [he] owes the State no Allegiance," apparently swayed Jefferson's opinion in favor of the defendant.

In 1779, John Tayloe II died, and left both ironworks to his son, John Tayloe III. Only eight years old at the time of his father's death, the younger Tayloe would play no part in his father's numerous enterprises until the 1790s. Thomas Lawson dutifully stayed on at Neabsco until his own death in 1785, although his "being rather infirm" prevented him from attending to the business in his usual competent manner.[146] Over the course of four decades, John Tayloe II had constructed a multifaceted industrial complex out of small rural iron furnace. Altogether, Tayloe founded and managed two furnaces, a forge, several profitable mills, an active shipbuilding enterprise, and merchant and landlord investments—an extremely valuable and profitable estate. The Occoquan Company alone was valued at over £20,000, and Tayloe's total holdings in Prince William County were likely worth at least double that amount. These industrial activities allowed Tayloe to reduce his dependence on the tobacco market while simultaneously shielding him from the worst effects of tobacco market fluctuation.

Endings and Beginnings under the Management of John Tayloe III

When John Tayloe III assumed management responsibilities for the entire Tayloe estate, he made efforts to maximize the profitability of his holdings. By the 1790s, both Neabsco and Occoquan were deteriorating physically and productively. The only remaining mention of the Occoquan estate during this time is through a series of negotiations between the third Tayloe and second Presley Thornton of Northumberland County. Tayloe's and Thornton's fathers had been the founding partners of the Occoquan Ironworks, and both men bequeathed their shares in the estate to their sons. Tayloe therefore owned three-quarters of the estate, while Thornton inherited the remaining one-quarter. During his lifetime, the elder Thornton had served mostly as a silent partner, only occasionally participating in the daily operations of the works. His son's desire to dispose of his small

Figure 13
John Tayloe III (1771–1828)

John Tayloe III recognized that ironmaking was on the decline in Prince William County because of depleted natural resources. He closed down the ironworks at Neabsco and Occoquan by 1800 and reoriented activities there towards agriculture, shipbuilding, and plantation service industries like shoemaking, blacksmithing, and carpentry. (Papers of the Randolph Family (MSS 10503), Special Collections Department, University of Virginia Library)

share seems to indicate that he had even less interest in being an ironmaster than his father. During the 1790s, Tayloe and Thornton arranged for Tayloe to become the sole owner of the entire works, which were valued at as much as £20,000. In return, Tayloe released Thornton from all debts then on the books owed by the estate.[147]

Throughout the 1790s, Tayloe's manager at Neabsco was William Holburne, who had begun his career with the Tayloes under John Tayloe II and replaced Thomas Lawson during the 1780s. Holburne oversaw the production of pig iron at Neabsco, handled all the marketing arrangements, and managed the labor of the numerous slave ironworkers and agricultural field hands. However, Holburne faced a serious problem in making Neabsco profitable and productive—an increasing deficiency in natural resources. Neabsco had been producing iron for over 60 years, and thus had also been steadily stripping the surrounding acres of the lumber required for making charcoal. U.S. Capitol Architect Benjamin Latrobe visited Neabsco during this period and described the state of the works as of 1796:

> It is not a work of any considerable Magnitude. Eight Hands by day and 8 by night are employed in it. It labours under this very great disadvantage, that all the Ore smelted in it is brought by Water from the Eastern shore of Maryland, to the Mouth of Occoquan bay [sic] and then carried by Waggons 2 Miles to the Furnace over very wretched roads. It is in blast only every third year. The woods which furnish Charcoal, already begin to feel a deficiency, although some thousand acres are appropriated to the Supply.[148]

While Tayloe and his managers had always been able to overcome the transportation-related problems Latrobe described, they could do little to halt the loss of the necessary natural resources.

In addition, Holburne was attempting to market his small quantities of pig iron in direct competition with the new western iron industry. Beginning during the revolutionary era, ironmasters increasingly moved west in search of better ore deposits and more plentiful natural resources. Once established, the older works such as Neabsco simply could not compete. Firms in western Virginia and Tennessee such as David Ross's Oxford Furnace and William Weaver's Buffalo Forge produced a large and steady stream of high-quality iron.[149] In his famous *Notes on the State of Virginia,* Thomas Jefferson commented on the productivity of the new western furnaces and forges: "Callaway's [Franklin County], Ross's [Campbell County], and Zane's [Frederick County], make about 150 tons of bar iron each, in the year. Ross's makes also about 1600 tons of pig iron annually; Ballendine's [Richmond City] 1000; Callaway's [Campbell County], Millar's [Augusta County], and Zane's, about 600 each."[150] Meanwhile, Tayloe regularly faced low prices at market, high supply and shipping costs, and a variety of resource deficiencies.

Tayloe was quick to face the new reality of his family's once-profitable ironworks. Between 1794 and 1800, he sold 11 tracts of Neabsco land totaling over 1,800 acres—more than one-third of the original 5,000-acre iron plantation.[151] By 1805, Tayloe decided to withdraw from Neabsco altogether. He advertised the property in the *Virginia Gazette* and other area newspapers: "*For Sale.* Neabsco Furnace And its appendages, with four or five thousands acres of LAND, adjoining, near the town of Dumfries, and within three or four miles of the Potomack river. . . . *A valuable* MINE BANK in Maryland, may be had with the Furnace, if required." Recognizing what remained valuable in the property, Tayloe pointed out that "THE soil is generally adapted to the produce of small grain."[152]

When he failed to find a buyer, Tayloe adjusted operations at Neabsco to take advantage of its remaining profitable features: land and slaves. Two activities dominated slave life at Neabsco in the years following the closing of the furnace: grain agriculture and craft trades. As he did on the Mt. Airy plantations in Richmond County, Tayloe cultivated the majority of his acreage at Neabsco in wheat, corn, and a variety of other grains. This produce supported the still large resident slave population, brought small profits at market, and also assisted in supplying the Tayloe family's needs when in residence at their new house in Washington, D.C., the Octagon, just 20 miles north of Neabsco. The latter purpose became even more important once the large Tayloe family moved permanently to Washington around 1817.[153] In addition, a majority of the slaves still plied a variety of craft trades at Neabsco. For obvious reasons, skilled slaves

Figure 14
Octagon House

"The City House" of John Tayloe III (1771–1828), constructed between 1799 and 1801 at the corner of 18th Street and New York Avenue in Washington, D.C. Tayloe lived in the house every year between April and October until 1816, and thereafter moved his family permanently to live in the city in order to conduct his business there and in northern Virginia. Tayloe's Prince William County plantations became particularly important with the construction of the Octagon because they supplied foodstuffs to the city house. (Photo by the author)

had always predominated in the population at the ironworks. Therefore, Neabsco steadily developed a reputation in Prince William County as a center for quality services and trades such as blacksmithing, carpentry, shipbuilding, and spinning and weaving. Slaves remained active in all of these areas, as well as others, even after iron production ceased. In 1824, for example, 92 slaves still resided at Neabsco, and held occupations including ship and house carpenters, sailors, blacksmiths, hostlers, millers, wagoners, weavers, spinners, cooks, laundresses, knitters, wheelwrights, woodcutters, coach hands, jobbers, hog minders, and house servants. The "Working Hands" or field hands numbered only 13 of the 92, and included two ploughmen and two cotton pickers. Eleven of the 92 were "Old" and were given only occasional small jobs so as "to be kept from idleness." Therefore, 74% (68/92) were skilled or semi-skilled and worked almost solely plying their craft or trade for Tayloe and neighborhood customers (see Table 3).[154]

In addition to the readjustments he made in Neabsco's purpose, Tayloe looked for other ways to maximize the profitability of his remaining Prince William County lands. During the 1790s, Tayloe built a home and purchased numerous lots in the new city of Washington, D.C. He began spending half of every year in the city after construction on his home was completed in 1801. Thereafter, Tayloe began investing in numerous projects characteristic of the post-revolutionary economy and needs of the new nation, including transportation, service industry, and internal improvements.

Tayloe established three hotels in Prince William County during the 1810s. Tayloe was operating the Farmers and Planters Hotel in Prince William as early as 1812, and leased the management of the property to Thomas and Edward Spence in December 1812. The Spences paid Tayloe $152.78 for nine months rent in 1812.[156] By June 1813, Tayloe turned to John Weathers to manage the hotel. Weathers agreed to pay Tayloe $300 per year during his five-year rental contract, as well as an additional £90 yearly in tavern expenses. Tayloe permitted Weathers the use of the surrounding houses, stables, and other outbuildings so long as Weathers returned them in at least the same condition as when he received them. In addition, Weathers agreed to keep the hotel "supplied with good meats, flour & meale, [and] also such Liquors *of good quality* as the times may offer & the calls of the House require."[157]

Tayloe operated two other small hotels both located in Prince William County. During the 1810s and 1820s, Tayloe operated the Occoquan Hotel in the town of Occoquan. Tayloe owned at

Table 3
Working Hands Belong to Neabsco, 1824[155]
(ages in parentheses)

Ship Carpenters
Reuben (56)
Natt (30)
Jim (36)
Sampson
Anthony (46)
Billy
Boys: Natt and Jim

House Carpenters
Kitt (52)
Lewis (47)
Scipio
Tom
Ned
Natt
John Garland
Boys: John and Lewis

Smiths
John (13)
Sam
George (29)
Tom
Alek
Davy Sailor (?) (29)

Shoe Makers
Henry
Jess

Hostlers
Ralph
Moses

Sailors in the *Henrietta*
Dick (37)
Dick
Gawen
Joe
Camunk (?)

Millers
Jason
Gawen

Waggoner
Sampson

Working Hands for Neabsco
(field hands)
Sawney (15)
Jim
Rose
Mary
Elleanor
Milly (58)
Judy
Old Cate
Arrabella

Ploughmen
Tom
Ralph

Weaver
Fanny (54)

Spinners
Becky (51)
Nelly (40)
Julia
Cate
Patt
Lissa
Lucy (44)
Sicilia

In the House
Letty
Tom
Joe

Cooks
Judy
Letty

Laundry
Moll

Knitter
Anna

Wheelwrights
Jacob (64)
Sampson (44)

"Old & to be employed in any way to be kept from idleness"
Arrabella
Mary
Pender
Harry
Scipio
Tony
Troy
Gawen

Woodcutters
Jim
Charles
Soloman
Bristol
Jack
Billy

Coach Hands
Ben
Jack

Jobbers
Billy
Morgan
George
Tom
Dick a boy
Old Godfrey

Hogminder
General

Cotton pickers
Hannah
Arrabella

Figure 15
Neabsco Inventory, 1825

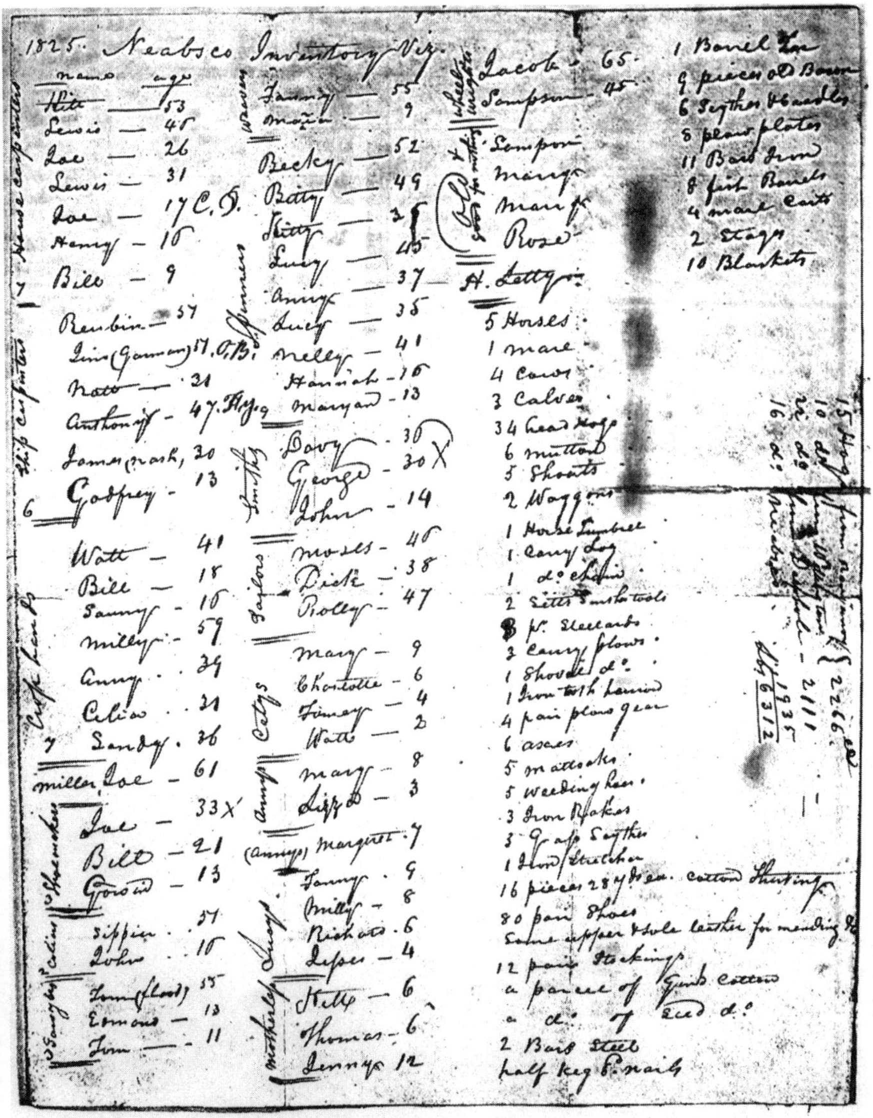

1825 Neabsco Inventory Viz.

name age
Hill — 53
Lewis — 45
Joe — 26
Lewis — 31
Joe — 17 C.S.
Henry — 15
Bill — 9
Reubin — 57
Jim (Gunman) 57 O.B.
Nate — 31
Anthony — 47 Fly.g
James (mark) 30
Godfrey — 13
Watt — 41
Bill — 18
Sammy — 15
Milly — 59
Anny — 39
Celia — 31
Sandy — 36
miller Joe — 61
Joe — 33
Bill — 21
Gowid — 13
Sippie — 57
John — 15
Tom (flood) 55
Edmond — 13
Tom — 11

Fanny — 55
Maria — 9
Becky — 52
Betty — 49
Kitty — 3
Lucy — 45
Anny — 37
Lucy — 35
Nelly — 41
Hannah — 15
Maryan — 13
Davy — 30
George — 30
John — 14
Moses — 45
Dick — 38
Rolly — 47
Mary — 9
Charlotte — 6
Fanny — 4
Watt — 2
Mary — 8
Lizzy — 3
Margret — 7
Fanny — 9
Milly — 8
Richard — 6
Lizzie — 4
Kelly — 6
Thomas — 6
Jenny — 12

Jacob — 65
Sampson — 45
Sampson
Mary
Mary
Rose
H. Letty

5 Horses
1 mare
4 Cows
3 Calves
34 head Hogs
6 mutton
5 Shoats
2 Waggons
1 Horse Tumbrel
1 Carry Log
1 do. Chain
2 Setts Smiths tools
3 pr. Steelyards
3 Carry plows
1 Shovel do.
1 Iron tooth harrow
4 pair plow gear
6 axes
5 mattocks
5 weeding hoes
3 Iron Rakes
3 Grass Scythes
1 Hand Stretcher
16 pieces 28 yds ea. cotton Sheeting
80 pair Shoes
Some upper & sole leather for mending
12 pair Stockings
a parcel of Gins cotton
a do. of Seed do.
2 Bars Steel
half keg 8d. nails

1 Barrel Tar
9 pieces old Bacon
6 Scythes & cradles
8 plow plates
11 Bars Iron
8 fish Barrels
4 mare Carts
2 Stags
10 Blankets

Even after ironmaking ceased at Neabsco, skilled slave labor remained central to the plantation. The 67 slaves listed here worked as house and ship carpenters, shoemakers, colliers, millers, sawyers, blacksmiths, spinners, and weavers, among other occupations. These men and women earned Tayloe's Prince William plantation a reputation as a center for high quality craft work. (Courtesy, Virginia Historical Society)

Figure 16
Map of Prince William County, John Wood, 1820

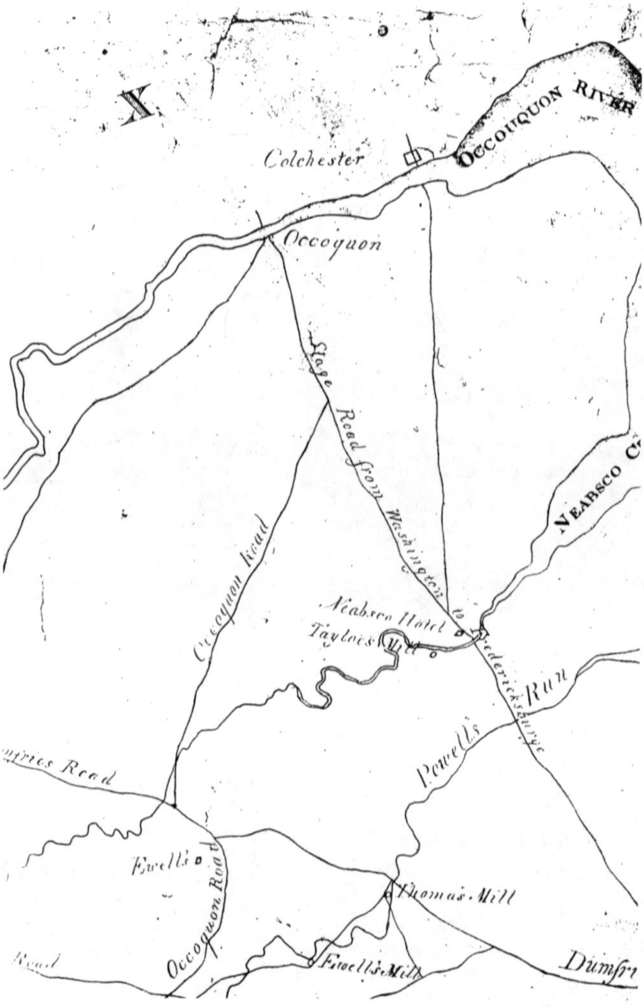

This map shows two of Tayloe's businesses in the years after he shut down iron production at Neabsco: milling and hostelry. Tayloe's mill indicates the location of the old furnace complex at Neabsco. Nearby, he opened the Neabsco Hotel along his stage coach route from Washington to Fredericksburg. Though Tayloe had moved to Washington in 1817, he remained heavily invested in Virginia businesses. (Courtesy, Library of Virginia)

least three lots in the town, numbers 21, 45, and 46.[158] Tayloe purchased all three lots during the spring and summer of 1814 from Robert Walsh and John Janney. Tayloe may also have inherited other lots from his father in connection with the family's old ironworks complex located on the outskirts of the town. There are no records as to whether these lots contained any improvements when Tayloe purchased them. However, by the end of the decade, Tayloe was operating the Occoquan Hotel on these properties.[159] Finally, Tayloe also owned a hotel on his Neabsco estate called the Neabsco Hotel. There are no records that reveal when a hotel or tavern might have been established at Neabsco. Undoubtedly, Tayloe's interest in operating hotels in Prince William County was related to his interest in offering transportation services. However, the only mention of the Neabsco Hotel occurred when Tayloe's sons sold it and 126 surrounding acres to Araminta and Zebulon Kankey in 1831 for $800.[160]

Tayloe developed his interest in offering transportation services in a circuitous manner. One of his many county government offices included that of postmaster for Richmond County.[161] Tayloe held this office roughly between 1800 and 1813. As he became more entrenched in the business world of Washington City, Tayloe began to hand off his postmaster responsibilities to a hand-picked successor.[162] By 1817, Tayloe moved his family permanently from his Mt. Airy Plantation to his city house. By 1818, Tayloe became involved in the postal business of the city and began operating several mail coaches. The first stageline Tayloe managed was the Dispatch & Expedition Stage that ran between Washington and Baltimore. In 1818, Tayloe made almost $2,000 carrying mail and passengers between Washington and Baltimore.[163]

In 1819, Tayloe contracted "to carry the mail from Washington via Georgetown to Fredericktown." When someone else also made a bid on this mail route, Tayloe offered to take only $500 in salary to carry the mail—$300 below that advertised. Tayloe went on to say that if the competitor matched his offer, he would do it *"for nothing*—sooner than not procure the Contract." Tayloe even indicated that he would be willing to pay for the contract.[164] Tayloe considered carrying the mail to be an extraordinary business opportunity as mail coaches, which ran on a particular and reliable schedule, also provided passenger transportation—the truly profitable end of the business. Tayloe won the Washington-Georgetown-Frederick contract, although its particular terms have not survived. In the first year that he operated both lines, Tayloe made $4,387.16, more than doubling his profit from the previous year.[165]

Within a year or two, Tayloe bought out his competition for the most prominent southern mail route. One Mr. P. Hewitt of Alexandria operated the line that ran from Washington City to Stafford Courthouse via Alexandria and Dumfries. Tayloe offered to buy his horses and stage and wanted Hewitt to not only agree not to attempt to procure the contract but also to do everything to help Tayloe get it. Once accomplished, this route in 1822 brought Tayloe a postal salary alone of $6,100, minus $912 in bridge tolls. Tayloe's interest in obtaining the southern mail route corresponds to his opening of the Occoquan and, likely, Neabsco Hotels. Tayloe often interchanged Occoquan and Dumfries as stopping points on the southern route. Tayloe realized that coach stops at his Prince William County hotels and taverns meant additional food and beverage revenue from his stagecoach passengers. By the mid-1820s, Tayloe decided to concentrate his energies in Occoquan solely on the stageline and sold his hotel there to Michael Cleary. In exchange for providing Tayloe and his family free accommodations at the hotel as long as he was in charge of the mail, Tayloe supplied Cleary and the hotel with 500 pickled shad from his Deep Hole plantation fishery, located in Prince William County between the Marumsco Creek and Occoquan River.[166] Tayloe made bids for other lines in Virginia as well, including the line from Stafford Courthouse west to Fredericksburg and from there on to Bowling Green. Tayloe anticipated that these two lines would bring him almost a $5,000 postal salary.[167]

Figure 17
Map of Botetourt County, 1821

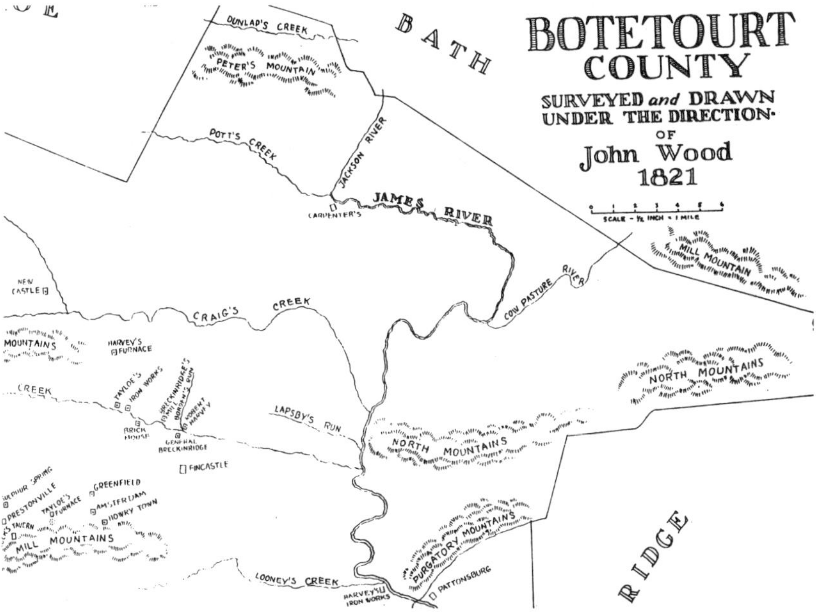

Tayloe remained interested in ironmaking despite its decline in the Tidewater. In 1810, an opportunity arose for him to participate in the lucrative western iron industry. That year, he purchased the Cloverdale Furnace and Brunswick Forge (bottom left of map) in Botetourt County and began sending some of his skilled Prince William slaves west. (Courtesy, Virginia Historical Society)

Even though Tayloe managed to maintain Neabsco's profitability in other ways, he remained interested in iron manufacturing and adjusted easily to the new realities of the industry. Tayloe recognized, as certainly anyone related to the industry did, that the new locus of the iron industry had shifted to the west. Tayloe therefore closely watched and waited for the perfect opportunity to make itself available. During the spring of 1810, just such an opportunity arose. Both a forge and a furnace became available on Tinker Creek in Botetourt County. James Breckinridge and Carter Beverly, the remaining partners in an

Figure 18
Map of Deep Hole Plantation

John Tayloe II purchased the lands for Deep Hole in 1768. His son, John III, was probably the one who founded the Fishery. By the nineteenth century, John III was supplying his Prince William taverns and hotels with supplies of fish from Deep Hole. (Courtesy, Virginia Historical Society)

ironworks that had been in operation as early as 1787, sold Tayloe the Cloverdale Furnace and Brunswick Forge. Tayloe paid Beverly £20,000 for the 13 tracts that comprised Brunswick, including a 187-acre "Ore Bank Tract." He paid Breckinridge a comparatively low £1,900 for Cloverdale's 751 acres.[168] Thereafter, Tayloe spent several years fine tuning the layout and com-

Figure 19
Map of Neabsco, 1820s

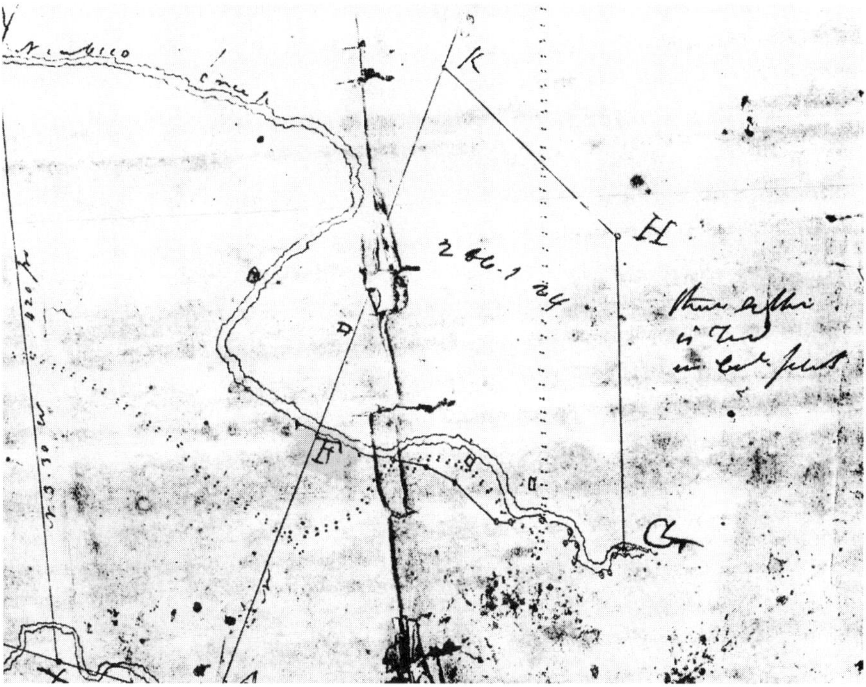

This "Map [of] Col. Jn Tayloe['s lands]" in Prince William County, marked the location of the old Neabsco Furnace. After Tayloe died in 1828, his executors sold most of his remaining Prince William County lands out of the Tayloe family. (Courtesy, Virginia Historical Society)

position of his Botetourt County estate, ultimately expanding it to include two additional works called Martha's Furnace and Tayloe's Slitting Mill Forge.[169] Tayloe began selling off his lands in Prince William County to fund his new purchases in the west and in Washington City.

After the third John Tayloe's death in 1828, his sons sold most of the remaining Neabsco lands. The year 1828 marked the end of large-scale Tayloe-family involvement in Prince William County. After a century of ironmaking in northern Virginia, Tayloe's descendants moved on to other enterprises and opportunities. The participation of these three generations of men in ironmaking in Prince William County represents two important facts about the business history of early Virginia. First, that the major Virginia planters were entrepreneurial and willing to engage in a variety of business ventures in addition to planting. Each Tayloe patriarch expressed his entrepreneurial bent in different ways that responded to specific social and economic factors that characterized their respective lifetimes. Each had his share of successes, failures, and near-misses but, more importantly, each attempted to make the most of what they had in order to secure a comfortable life for themselves and their heirs. Each produced for the market and in response to changes within it. Second, slave labor served as the lubricant in the machine of entrepreneurship. The presence of large populations of enslaved African Americans enabled the Tayloes and other planters to branch out from staple agriculture and ultimately necessitated that they continue to do so. Slaves demonstrated their abilities, became central to the daily operations of the South's business culture, and made the enterprises planters founded profitable. The Tayloes' Prince William County iron plantations would not have been as successful nor as enduring without the sweat and skills of their African-American slaves.

Between the 1730s and 1830s, three generations of the Tayloe family played a significant role in laying the foundations of business and community in Prince William County. Their names are not memorialized in the landscape of the county as, for example, the Lees are. But the Tayloes nonetheless were one of the founding families of the county and were responsible for helping to create a regional economy in the eighteenth century that tied Prince William together with the Northern Neck of Virginia, Washington and Baltimore, and beyond.

Endnotes

1. Lorena S. Walsh, "Plantation Management in the Chesapeake," *Journal of Economic History* 49 (1989), 394.

2. *Ibid;* Christine Daniels, "Gresham's Laws: Labor Management on an Early-Eighteenth-Century Chesapeake Plantation," *Journal of Southern History* (hereafter *JSH*) 62 (May 1996), 208–209; Richard L. Bushman, "Markets and Composite Farms in Early America," *William and Mary Quarterly* (hereafter *WMQ*) 3d ser., 55 (3) (July 1998), 356; James B. Slaughter, *Settlers, Southerners, Americans: The History of Essex County, Virginia, 1608–1984* (Salem, W.V., 1985), 14; Russell R. Menard, "The Tobacco Industry in the Chesapeake Colonies, 1617–1730," *Research in Economic History* V (1980), 114, 125–126; Allan Kulikoff, *Tobacco and Slaves: The Development of Southern Cultures in the Chesapeake, 1680–1800* (Chapel Hill, 1986), 78; John J. McCusker and Russell R. Menard, *The Economy of British America, 1607–1789* (Chapel Hill, 1985), 119; Aubrey C. Land, *Bases of the Plantation Society* (Columbia, S.C., 1969), 99, 101; Gregory Stiverson, "'Gentlemen of Industry, Skill, and Application': Plantation Management in Eighteenth Century Virginia," (Williamsburg, VA, 1975), 97–98; Carville V. Earle, *The Evolution of a Tidewater Settlement System: All Hallow's Parish, Maryland, 1650–1783* (Chicago, 1975), 131–132.

3. Carr and Walsh, "Economic Diversification and Labor Organization," 145–146, 163; Sheridan, "The Domestic Economy," 46–47; Land, *Bases of the Plantation Society,* 101, 125; Paul G. E. Clemens, "The Operation of an Eighteenth-Century Chesapeake Tobacco Plantation," *Agricultural History* 44 (July 1975), 518, 521–522; McCusker and Menard, *The Economy of British America,* 127; Jean B. Russo, "A Model Planter: Edward Lloyd IV of Maryland, 1770–1796," *WMQ* 3d ser., 49 (January 1992), 62.

4. Joyce E. Chaplin, *An Anxious Pursuit: Agricultural Innovation and Modernity in the Lower South, 1730–1815* (Chapel Hill, 1993), 187; see

also, Carr and Walsh, "Economic Diversification and Labor Organization," 146; Clemens, "The Operation of an Eighteenth-Century Chesapeake Tobacco Plantation," 525.

5. Richmond County Order Book, 6: 29, 178, 258–259, 228, 470; 7: 78; 8: 234, 9: 59, 151; Richmond County Deed Book 6: 141; 7: 18–19; 8: 61–63, 136; TFP, 5: 96–97; W. Noel Sainsburg, ed., *Calendar of State Papers, Colonial Series, America and the West Indies* (hereafter *CSP*) (Vaduz, 1964), 38: 387; "The Colonial Council of Virginia," *WMQ,* 1st ser., 3 (1) (July 1894), 67, this articles mistakenly identifies 1734 instead of 1732 as Tayloe's first year on the Council; H. R. McIlwaine, ed., *Executive Journals of the Council of Virginia* (hereafter *Executive Journals*) (Richmond, 1925), 4: 336, 375; Charles Kemper, ed., "The Early Westward Movement of Virginia, 1722–1734, as Shown by the Proceedings of the Colonial Council," *Virginia Magazine of History and Biography* (hereafter *VMHB*) 13 (4) (April 1906), 360. See, Rhys Isaac, *The Transformation of Virginia, 1740–1790* (New York, 1982), 133–135, for a good discussion of the relationship between power and access to wealth in early Virginia; Rudolf Braun, "The Rise of a Rural Class of Entrepreneurs," *Journal of World History* 10 (1967), 554.

6. King George County Deed Book, 1: 51–52; Essex County Deed Book, 17: 174. See also, G. McClaren Brydon, "The Bristol Iron Works in King George County," *VMHB* 42 (2) (April 1934), 97–102; Edward Heite, "The Pioneer Phase of the Chesapeake Iron Industry: Naturalization of a Technology," *Quarterly Bulletin of the Archaeological Society of Virginia* 38 (3) (1983), 144.

7. Hening's *Statutes,* 1: 488, 2: 124–125, 493–494. These acts, passed in 1657/8, 1661/2, and 1682, declared that the "exportation of hides, wool, and iron is not allowed" and was punishable by fines dependent upon the amount exported. Mercantilist theory held that colonies stimulated trade by serving as outlets for English manufactured goods and providing raw materials in return. Practically, this meant that England restricted all trade with other imperial powers, especially the Dutch during the seventeenth century, and the processing of raw materials into manufactured goods in the colonies themselves. These restrictions were codified in the Navigation Acts, which largely lacked enforcement mechanisms.

8. Michael W. Robbins, *The Principio Company: Iron-Making in Colonial Maryland, 1720–1781* (New York, 1986), 11–12; Heite, "The Pioneer Phase of the Chesapeake Iron Industry," 139.

9. Hening's *Statutes,* 4: 228–231. Similar acts were levied in 1730 and 1748, *Ibid.,* 4: 296–299, 6: 137–140. See also, C. Robert Haywood, "Eco-

nomic Sanctions: Use of the Threat of Manufacturing by the Southern Colonies," *JSH* 25 (2) (May 1959), 210–211.

10. John T. Schlotterbeck, "Plantation and Farm: Social and Economic Change in Orange and Green Counties, Virginia, 1716–1860," unpublished dissertation, Johns Hopkins University, 1980; Heite, "The Pioneer Phase of the Chesapeake Iron Industry," 142–143.

11. Keach Johnson, "Genesis of the Baltimore Company," *JSH* 19 (May 1953), 157–179, quote, 162; Robbins, *The Principio Company, passim;* Ronald Lewis, *Coal, Iron, and Slaves: Industrial Slavery in Maryland and Virginia, 1715–1865* (Westport, Conn., 1979), chapter 1.

12. King George County Deed Book, 1: 14–19, 641–642, 1A: 115–118, 2: 59–63; Peggy Joyner, compiler, *Abstracts of Virginia's Northern Neck Warrants, 1710–1780* (hereafter *Northern Neck Abstracts*) (Portsmouth, Va., 1986), 153.

13. Richmond County Deed Book, 8: 488–491.

14. King George County Deed Book, 2: 429–431.

15. Account of John Tayloe [with the Bristol Ironworks], Tayloe Family Papers, VHS, Reel 1, Frame 38. A large part of the Tayloe Family Papers at the VHS have been microfilmed as part of the *Records of Ante-Bellum Southern Plantations,* Series M, Part 1: The Tayloe Family. These papers will hereafter be expressed as TFP, Reel: Frame, to distinguish them from Tayloe family collections that are not part of the microfilm collection located at the VHS and other repositories.

16. King George County Deed Book, 2: 445–446.

17. Robert Carter to Colo. John Tayloe, January 29, 1728/9, Robert Carter Letterbook, 1728–1730, VHS; Louis Morton, *Robert Carter of Nomini Hall: A Virginia Planter of the Eighteenth Century* (Charlottesville, Va., 1965), 18.

18. Stafford County Deeds, TFP, 5: 39, 41; Prince William County Deed Book, A, 153, 387; B: 4–7, 31, 34, 245, 439; D: 364; E: 10; See also, Deeds, TFP, 5: 53, 602, 604.

19. *Northern Neck Abstracts,* ix–xii; Hughes, *Surveyors and Statesmen,* 107.

20. No deed evidence exists to pinpoint the precise date of purchase. However, his appeal to the Council to allow duty-free importation of ore in 1738, and his 1741 account book entries identifying his ore source as his "new Bank" point to this three-year period. By the 1750s, John Tayloe II owned about 300 acres in Anne Arundel and Baltimore Counties, from which he mined iron ore for the use of his Vir-

ginia ironworks. These lands were likely the ones purchased by John Tayloe I. Baltimore County Certificates, Patented, May 7, 1757, Maryland State Archives, S1190-65, Folder 4818; Account Book, TFP, 1: 38.

21. *Executive Journals*, 4: 433; William P. Palmer, ed., *Calendar of Virginia State papers and Other Manuscripts* (hereafter *CVSP*) (New York, 1968), 38: 387.

22. John Tayloe I Account Book, TFP, 1: 5–45.

23. Kathleen Bruce, *Virginia Iron Manufactures in the Slave Era* (New York, 1968), 18.

24. "Ore Receipts," 1755–1757, TFP, 56: 409.

25. John Tayloe I Account Book, 1740–1741, TFP, 1: 35.

26. Carter L. Hudgins and Douglas W. Sanford, "Neabsco Mills Ironworks Site, National Register of Historic Places Registration Form," Center for Historic Preservation, Mary Washington College, 1993, Section 7, 1–2. A copy of this form is also located in the Neabsco Furnace File, Tate Thompson Brady Papers, VHS. Only Phase I and II archaeological investigations have been conducted at the Neabsco Ironworks Site (44PW629), which is located on private property. The preliminary fieldwork indicated that the site maintained a high degree of archaeological integrity.

27. Hudgins and Sanford, "Neabsco Mills Ironworks Site," Section 7, 3–4.

28. John Tayloe I Account Book, 1740–1741, TFP, 1: 23–25; John Tayloe Estate Inventory.

29. John Tayloe I Account Book, 1740–1741, TFP, 1: 5–45.

30. *Ibid.;* Will of John Tayloe I, 1747, TFP, 5: 103–112. Sloops and schooners were among the smallest, quickest, and most frequently built ships in the colonies. Sloops averaged 31 tons while schooners averaged 34 tons. See, William Kelso, "Shipbuilding in Virginia, 1763–1774," *Records of the Columbia Historical Society* (1973), 2–4; Arthur Pierce Middleton, "Ships and Shipbuilding in the Chesapeake Bay and Tributaries," in Ernest M. Eller, ed., *The Chesapeake Bay in the American Revolution* (Centerville, Md., 1981), 98.

31. John Tayloe I Account Book, 1740–1741, TFP, 1: 31.

32. Tayloe did purchase two slaves who had worked at Bristol for his Neabsco Ironworks. John Tayloe I Account Book, 1740–1741, TFP, 1: 5–45.

33. Allan Kulikoff, *Tobacco and Slaves: The Development of Southern Cultures in the Chesapeake, 1680–1800* (Chapel Hill, 1986), 68, 70, 355–356; John Tayloe Estate Inventory.

34. Travers Nash Deposition, March 1789, File 171. John Tayloe Estate Inventory. The 1747 inventory includes 100 slaves (69 men, 31 women) and seems to indicate that Nash underestimated the number of slaves working at Neabsco in 1755.

35. By the 1750s, Tayloe owned about 300 acres on the Patapsco River in Anne Arundel and Baltimore Counties, Maryland, from which he mined iron ore for the use of his Virginia ironworks. It is likely that his father had purchased these lands during the 1730s, as he petitioned the Virginia Council to allow him to import Maryland ore duty-free in 1737, however no deed evidence exists to pinpoint the precise date of purchase. In 1757, Tayloe had his lands resurveyed in order to combine three contiguous tracts into one piece of property named "Tayloe's Lot," which served as his source of iron ore. Baltimore County Certificates, Patented, May 7, 1757, Maryland State Archives (hereafter MSA), S1190-65, Folder 4818.

36. *Ibid.*

37. John Tayloe Estate Inventory; Charles Chiswell as quoted in Robbins, *The Principio Company*, 92.

38. Robbins, *The Principio Company*, 121–122; "Scheme of a Lottery, for disposing of certain lands," *Virginia Gazette*, December 1, 1768.

39. Will of John Tayloe I.

40. Robbins, *The Principio Company*, 3, 18; Keach Johnson, "The Baltimore Company Seeks English Markets: A Study of the Anglo-American Iron Trade, 1731–1755," *WMQ* 3d ser., 16 (1959), 40; David Curtis Skaggs, "John Semple and the Development of the Potomac River Valley, 1750–1773," *VMHB* 92 (3) (1984), 284–285.

41. Robbins, *Principio Company*, 193–195; Johnson, "Baltimore Company," 39n9; Arthur Cecil Bining, *British Regulation of the Colonial Iron Industry* (Clifton, N.J., 1973), 128–133; Lewis, *Coal, Iron, and Slaves*, 11–12.

42. Robbins, *Principio Company*, 193–195.

43. Lewis, *Coal, Iron, and Slaves*, 11; Skaggs, "John Semple," 284; McCusker and Menard, *The Economy of British America*, 130–132. The Iron Act of 1750 allowed pig and bar iron to be shipped duty-free to London, and was later expanded to include all English ports. However, the colonies were not permitted to manufacture iron into steel or other iron products. It is possible such restrictions on the development of planter-businessmen's entrepreneurial interests helped predispose them to Revolution. Robbins, *Principio Company*, 13.

44. These figures were drawn from tables in Robbins, *Principio Company*, 193–195, and Bining, *British Regulation*, 128–133.

45. Tonnage rounded to the nearest ton.

46. By the middle of the eighteenth century, other colonies, such as Pennsylvania and Massachusetts, were catching up on the Chesapeake in terms of iron production.

47. Skaggs, "John Semple," 285; Heite, "Pioneer Phase of the Chesapeake Iron Industry," 140; Robbins, *Principio Company,* 6.

48. Robbins, *Principio Company,* 6–8; Keach Johnson, "Genesis of the Baltimore Company," *JSH* 19 (May 1953), 166; Heite, "Pioneer Phase of the Chesapeake Iron Industry," 141; William Reynolds, "An Account of the Albemarle Iron Works," *Albemarle County History* 50 (1992), 41.

49. Robbins, *Principio Company,* 8–9, 20–21; Heite, "Pioneer Phase of the Chesapeake Iron Industry, 141; "An account of the sizes of IRON proper to be imported from America for his Majesty's Service," *Virginia Gazette,* April 27, 1769.

50. John Ballendine to John Tayloe, May 13, 1756, TFP, 54: 1034–1035; Tayloe and Thornton Land Lists, TFP, 5: 153; Prince William County Deed Book P: 201–210. See also, Skaggs, "John Semple," 288.

51. *Maryland Gazette,* November 25, 1756. See also, Skaggs, "John Semple," 289.

52. Prince William County Deed Book P: 7–8, 299–308; Tayloe and Thornton Land Lists, TFP, 5: 153; Anne Arundel County Deed Book BB 2, 629–631, MSA Land Records. See also, Skaggs, "John Semple," 289–290.

53. Archdeacon Burnaby as quoted in Fairfax Harrison, *Landmarks of Old Prince William: A Study of Origins in Northern Virginia* (Baltimore, 1987), 2: 428.

54. *Virginia Gazette,* June 13, 1766; Skaggs, "John Semple," 290. It is possible that others became embroiled in the conflict. In 1769, Thomas Tabb and Thomas Yuille placed a similar notice, declaring that they had purchased land and slaves from Ballendine in Fauquier, Prince William, and Fairfax Counties, and forewarning "all persons from making any purchase of the said lands or slaves." The slaves included a number of men with ironmaking skills. *Virginia Gazette,* November 16, 1769.

55. Thomas Lawson, Neabsco, to John Tayloe, October 27, 1771, File 171.

56. *Virginia Herald and Fredericksburg Advertiser,* June 5, 1788.

57. John Tayloe, Mt. Airy, to William Lee, Merchant at London, June 12, 1771, William Lee Papers, VHS.

58. The archaeological sites of both the furnace and forge have been identified. The furnace remains (44PW605) are located on the present county reservoir property immediately west of Occoquan. No above-ground remains of the forge (44PW606) are visible, but the documentary evidence proves its location was in the old mill complex at the west end of the town. No excavations have been conducted at either site.

59. *Virginia Herald and Fredericksburg Advertiser,* June 5, 1788; Occoquan Forge and Furnace Files, Tate Thompson Brady Papers, VHS.

60. While the evidence is only circumstantial, it is likely that the house referred to in these documents is Rockledge, which still stands in Occoquan overlooking the Occoquan River adjacent to the old forge site.

61. Occoquan Company Inventory, March 4, 1771, Colonial Williamsburg Foundation Library (hereafter CWF).

62. The following discussion of the business operations of Neabsco and Occoquan is based on the following (often incomplete) account books: John Tayloe II Account Book, 1749–1768, TFP, 2: 179–214; John Tayloe II Ledger, 1747–1787, Library of Virginia (hereafter, LVA; this is a copy of the 1749–1768 account book with some additional entries made by Tayloe's executors); Accounts, Bonds, Orders, 1756–1762, TFP, 56: 392–427; John Tayloe II Account Book, 1770–1776, Tayloe Family Papers, Mss1T2118g1, VHS; and, Thomas Lawson's Occoquan Accounts, 1757–1785, File 171. These account books reference many other account books which apparently no longer exist.

63. For more information about eighteenth-century accounting practices, see John Mair, *Book-Keeping Moderniz'd; or, A Methodical Treatise of Merchant-Accompts; or, Merchant-Accompts by Double Entry* (Edinburgh, 1736, 1773); W. T. Baxter, "Accounting in Colonial America," in A. C. Littleton and B. S. Yamey, eds., *Studies in the History of Accounting* (Homewood, Ill., 1956), 272–287; Albert F. Voke, "Accounting Methods of Colonial Merchants in Virginia," *Journal of Accountancy* 41 (7) (July 1926), 1–11.

64. Richard Parker's Deposition, March 20, 1790, File 171.

65. Correspondence and other evidence indicates that Neabsco produced pig iron from the late 1730s until the turn of the century and that Occoquan produced pig iron from about 1755 until the early 1780s. This discussion is then limited by the scope of existing account books, see note 31.

66. Occoquan Company Accounts, TFP, 2: 186, 196, 212; Handman

Lancaster Accounts in John Tayloe II Account Book, 1749–1768; Receipts, TFP, 56: 394, 400;

67. John Tayloe, Mt. Airy, Virginia, to William Lee, Merchant at London, June 12, 1771, William Lee Papers, VHS. By 1772, Tayloe and Russell had apparently come to some agreement, as Tayloe was again shipping his tobacco and iron to Russell.

68. Kulikoff, *Tobacco and Slaves,* 119; McCusker and Menard, *The Economy of British America,* 62–63, 121; Ronald Hoffman, *A Spirit of Dissension: Economics, Politics, and the Revolution in Maryland* (Baltimore, 1973), 18.

69. See McCusker and Menard, *The Economy of British America,* 119, 126–127, for a discussion of motivations for and the impact of diversification on the tobacco economy.

70. These figures for iron exportation from Neabsco and Occoquan are based on the following (often incomplete) account books: John Tayloe II Account Book, 1749–1768, TFP, 2: 179–214; John Tayloe II Ledger, 1747–1787, Library of Virginia (hereafter, LVA; this is a copy of the 1749–1768 account book with some additional entries made by Tayloe's executors); Accounts, Bonds, Orders, 1756–1762, TFP, 56: 392–427; John Tayloe II Account Book, 1770–1776, Tayloe Family Papers, Mss1T2118g1, VHS; and, Thomas Lawson's Occoquan Accounts, 1757–1785, File 171. These account books reference many other account books which apparently no longer exist. Pounds sterling amounts have been rounded to the nearest pound.

71. William Byrd II, "Progress to the Mines," in Louis B. Wright, ed., *Prose Works; Narratives of a Colonial Virginian* (Cambridge, Mass., 1966), 347–348.

72. Johnson, "Baltimore Company," 47; Johnson, "Genesis of the Baltimore Company," 157, 169–170.

73. Johnson, "Baltimore Company," 56–58.

74. Principio Company File, Tate Thompson Brady Papers, VHS. See also, Robbins, *Principio Company,* 56.

75. William Peden, ed., *Notes on the State of Virginia,* by Thomas Jefferson (New York, 1982), 27–28.

76. Johnson, "Baltimore Company," 56–58.

77. *Ibid.*

78. "A List of Lands," TFP, 5: 152–153; Prince William County Deed Book Q: 196, 205–208, 209, 356, 374–376; *Northern Neck Abstracts,* 129.

79. Occoquan Company Accounts, TFP, 2: 186, 196; Presly Thornton Account with John Alexander, February 18, 1759, TFP, 55: 404; Thomas Lawson's Occoquan Accounts, 1757–1785, File 171; Occoquan Furnace Inventory; Thomas Lawson, Neabsco, to John Tayloe, Christmas Eve, 1771, File 171.

80. Travers Nash Deposition, March 1789, File 171. See also, "The Miller in Eighteenth-Century Virginia: An Account of Mills & the Craft of Milling, as well as a Description of the Windmill near the Palace in Williamsburg," (Williamsburg, Va., 1990), (10–13).

81. McCusker and Menard, *The Economy of British America*, 323–325; Thomas Berry, "The Rise of Flour Milling in Richmond," *VMHB* 18 (October 1970), 390. All of these authors agree that milling was an industry that required entrepreneurial talent. They define mills as the basis of early industrialization in British America, reinforcing my argument that the North and South were economically, developmentally, and technologically similar before the 1790s.

82. John Ballendine to John Tayloe, May 13, 1756, TFP, 54: 1034–1035.

83. Occoquan Company Accounts, TFP, 2: 211; Special Report by the Commissioners, File 171; William Holburne Deposition, August 28, 1798, File 171.

84. Thomas Lawson, Neabsco, to John Tayloe, December 2, 1771, File 171.

85. Thomas Lawson, Neabsco, to John Tayloe, September 24, 1775, File 171.

86. Travers Nash Deposition, March 1789, File 171.

87. Thomas Lawson, Neabsco, to John Tayloe, October 27, 1771, File 171.

88. William Holburne Deposition, March 10, 1798, File 171; Occoquan Furnace Inventory; Thomas Lawson, Neabsco, to John Tayloe, Christmas Eve, 1771, File 171; Occoquan Company Accounts, TFP, 2: 211.

89. John Tayloe II Account Book, 1749–1768, makes numerous mentions of Tayloe's ships. See also Thomas Lawson, Neabsco, to John Tayloe, October 27, 1771, December 2, 1771, Christmas Eve, 1771, September 24, 1775, and November 20, 1777, File 171; William Lawson Deposition, March 11, 1789, File 171; William Holburne Deposition, May 10, 1798, File 171; Travers Nash Deposition, March 1789, File 171; Occoquan Company Inventory. Sloops and schooners were among the

smallest, quickest, and most frequently built ships in the colonies. Sloops averaged 31 tons while schooners averaged 34 tons. Brigs, ships, and snows were larger and slower vessels intended for the trans-Atlantic trade. See, Kelso, "Shipbuilding in Virginia," 2–4; Middleton, "Ships and Shipbuilding in the Chesapeake Bay and Tributaries," 98.

90. Kelso argued that ownership by a single man of more than four vessels comprised a shipline, and found that four such lines existed in Virginia during the Revolution (John Goodrich and John Greenwood both of Norfolk, John Thompson of Surry, and John Wilkins of Princess Anne County). See, "Shipbuilding in Virginia," 9.

91. Jacob Price, "Economic Function and the Growth of American Port Towns in the Eighteenth Century," *Perspectives in American History* 8 (1974), 168, 185; Middleton, "Ships and Shipbuilding in the Chesapeake Bay," 101, 108, 112, 120.

92. John Ballendine to John Tayloe, May 13, 1756.

93. Thomas Lawson, Neabsco, to John Tayloe, December 2, 1771, Christmas Eve, 1771, File 171.

94. *Ibid.*

95. Travers Nash Deposition, March 1789, File 171.

96. Anne Arundel County Deed Book BB 2, 629–631, MSA; Baltimore County Certificates, Patented, May 7, 1757, Folded 4818, MSA.

97. Thomas Lawson, Neabsco, to John Tayloe, December 2, 1771, File 171.

98. *Ibid.*

99. *Ibid.*

100. *Ibid.*

101. Thomas Lawson, Neabsco, to John Tayloe, December 24, 1771, September 24, 1775, File 171.

102. Travers Nash Deposition, March 1789, File 171.

103. Thomas Lawson Allison Deposition, August 7, 1798, File 171; William Lawson Deposition, March 11, 1789, File 171; William Holburne Deposition, May 10, 1798, File 171. Allison and Lawson were Thomas Lawson's relatives and parties to the countersuit. William Holburne became one of John Tayloe III's chief agents during the early nineteenth century.

104. Thomas Lawson, Neabsco, to John Tayloe, Christmas Eve 1771, November 20, 1777, File 171.

105. Thomas Lawson, Neabsco, to John Tayloe, November 20, 1777, File 171; William Lawson Deposition, March 11, 1789, File 171.

106. Occoquan Company Accounts, TFP, 2: 186; Stephen Innes, ed., *Work and Labor in Early America* (Chapel Hill, 1988), 10; Richard S. Dunn, "Servants and Slaves: The Recruitment and Employment of Labor," in Jack P. Greene and J. R. Pole, eds., *Colonial British America: Essays in the New History of the Early Modern Era* (Baltimore, 1984), 170–171.

107. *Maryland Gazette,* September 16, 1762; July 12, 1776.

108. *Maryland Gazette,* April 4, 1765; Occoquan Company Inventory, CWF.

109. Lewis, *Coal, Iron, and Slaves,* 131; Gerald Mullin, *Flight and Rebellion: Slave Resistance in Eighteenth-Century Virginia* (London, 1972), 83–98.

110. *Maryland Gazette,* October 10, 1754; September 6, 1759.

111. TFP, 56: 402; Westmoreland County Orders, February 25, 1760, 1758–1761, p. 103.

112. TFP, 56: 406; *Maryland Gazette,* April 4, 1765.

113. *Virginia Gazette,* August 4, 1768. See also, Mullin, *Flight and Rebellion,* 111.

114. *Virginia Gazette,* February 9, 1769.

115. *Ibid.*

116. Thomas Lawson's Occoquan Accounts, File 171; TFP, 56: 404.

117. William Lawson Deposition, March 11, 1789, File 171.

118. Occoquan Company Inventory, CWF. Only one eighteenth century inventory exists from Neabsco, although it does not provide details on slave occupations. See, John Tayloe Estate Inventory. However, several nineteenth-century inventories exist and are informative regarding slave occupations and the sexual division of labor at the ironworks. See, "Working Hands Belonging to Neabsco," 1824, TFP, 27: 773; 1825 Neabsco Inventory, TFP, 6: 396; 1828 Neabsco Inventory, TFP, 7: 272–276. See also, Robbins, *Principio Company,* 91–92; Samuel Sydney Bradford, "The Negro Ironworker in Ante-Bellum Manufacturing," *JSH* 25 (May 1959), 194–195; Charles B. Dew, "David Ross and the Oxford Iron Works: A Study of Industrial Slavery in the Early Nineteenth-Century South," *WMQ* 3d ser., 31 (April 1974), 195.

119. William Lawson Deposition, March 11, 1789, File 171; Thomas Lawson, Neabsco, to John Tayloe, Christmas Eve, 1771, File 171.

120. Thomas Lawson, Neabsco, to John Tayloe, Christmas Eve, 1771, File 171.

121. Thomas Jefferson's skilled slaves similarly led relatively independent and unsupervised lives and received preferential treatment because of their skills. See Lucia Stanton, *Slavery at Monticello* (Richmond, 1996), 27, 30.

122. Occoquan Company Inventory, CWF. For good discussions of slave naming practices, see, Gary B. Nash, *Forging Freedom: The Formation of Philadelphia's Black Community, 1720–1840* (Cambridge, Mass., 1988), 79–88; Herbert G. Gutman, *The Black Family in Slavery and Freedom, 1750–1925* (New York, 1976), 185–203; and, Peter H. Wood, *Black Majority: Negroes in Colonial South Carolina from 1670 through the Stono Rebellion* (New York, 1974), 181–186.

123. Occoquan Company Inventory, CWF; John Tayloe Estate Inventory; Gutman, *The Black Family in Slavery and Freedom,* 191–192.

124. Sidney W. Mintz and Richard Price, *The Birth of African-American Culture: An Anthropological Perspective* (Boston, 1992, 1976), 38–41; Kulikoff, *Tobacco and Slaves,* 318–319, 339–351; Lawrence W. Levine, *Black Culture and Black Consciousness: Afro-American Folk Thought from Slavery to Freedom* (Oxford, 1977), 4–5, 135; Brenda E. Stevenson, *Life in Black and White: Family and Community in the Slave South* (New York, 1996), 171–172.

125. See Table 2. This amount does not include the 104 tons produced in 1777 nor the 189 tons produced in 1778, as no value was recorded for production in those years.

126. Thomas Lawson, Neabsco, to John Tayloe, October 27, 1771, File 171.

127. Thomas Lawson, Neabsco, to John Tayloe, Christmas Eve, 1771, File 171; Thomas Lawson, Neabsco, to John Tayloe, September 24, 1775, File 171.

128. *Ibid.;* "Special Report by the Commissioners," File 171; William Holburne Deposition, August 28, 1798, File 171.

129. Thomas Lawson, Neabsco, to John Tayloe, October 27, 1771, File 171.

130. Thomas Lawson, Neabsco, to John Tayloe, September 24, 1775, File 171.

131. Thomas Lawson, Neabsco, to John Tayloe, October 27, 1771, File 171.

132. John Tayloe, Mt. Airy, to William Lee, Esquire, Merchant in London, June 15, 1773, William Lee Papers, VHS.

133. Thomas Lawson, Neabsco, to John Tayloe, September 24, 1775, File 171.

134. John Tayloe, Mt. Airy, to William Lee, Esquire, Merchant in London, March 25, 1774, William Lee Papers, VHS.

135. See Table 2 above.

136. John Tayloe, Mt. Airy, to William Lee, Esquire, Merchant in London, June 28, 1772, William Lee Papers, VHS.

137. Thomas Lawson, Neabsco, to John Tayloe, September 24, 1775, File 171.

138. *CVSP,* 8: 195; "Virginia Militias in the Revolution," *VMHB* 15 (June 1908), 92.

139. William Lawson Deposition, March 11, 1789, File 171.

140. Thomas Lawson, Neabsco, to John Tayloe, September 24, 1775, File 171.

141. Thomas Lawson to John Tayloe, November 20, 1777, File 171; William Lawson Deposition, March 11, 1789, File 171. For a good discussion of the impact of the Revolution on Virginia's economy, see, Sylvia R. Frey, *Water from the Rock: Black Resistance in a Revolutionary Age* (Princeton, 1991), Chapter 5.

142. Henry Lee to Governor Thomas Jefferson, April 9, 1781, in *CVSP,* 2: 21–23.

143. "Special Report by the Commissioners," File 171.

144. *Ibid.* Thomas Jefferson estimated that Virginia lost 30,000 slaves in 1781 alone. See Frey, *Water from the Rock,* 211 n. 22.

145. Frey, *Water from the Rock,* 143–152, 210–211.

146. Will of John Tayloe II, 1779, TFP, 5: 168–172; William Holburne Deposition, May 10, 1798, File 171; Richard Parker Deposition, March 20, 1798, File 171.

147. Fredericksburg Court Records, File 171. The documents in this file were created as part of a court battle between Tayloe and Thomas Lawson's descendants. After making this agreement, Tayloe discovered that Lawson had neglected to include his salary in the accounts for thirty years. Therefore, the outstanding book debts were actually much more significant than Tayloe had known when he struck the bargain with Thornton. The estates of John Tayloe II and Thomas Lawson thus instituted suits against each other over these debts, which resulted in an extremely rich set of sources on both Neabsco and Occoquan.

148. Edward C. Carter II, ed., *The Virginia Journals of Benjamin Henry Latrobe, 1795–1798* (New Haven, 1977), 1: 174.

149. See, for example, Charles Dew, *Bond of Iron: Master and Slave at Buffalo Forge* (New York, 1994); Dew, "David Ross and the Oxford Iron Works."

150. *Notes on the State of Virginia,* 27–28.

151. Prince William County Deed Book Y: 484, 618, 621, 623, 732; Z: 112; 1: 57; 2: 306; TFP, 5: 602

152. *The Virginia Gazette and Daily Advertiser,* August 10, 1805.

153. For an idea of the variety of articles the Tayloes had shipped from their various plantations to Washington, see Minute Book, 1813–1818, Southern Historical Collection, M1-4485; "A list of Articles for Mrs. Tayloes City House in Washington Sent from the Rappahannock Farms by Capt. Haney," November 1828, TFP, 27: 513.

154. "Working Hands Belonging to Neabsco," 1824, TFP, 27: 773.

155. Transcription of *Ibid.* When possible, ages were taken from 1825 Neabsco Inventory, TFP, 27: 673.

156. John Tayloe III Account Book, 1810–1814, TFP, 1: 315.

157. Contract, June 27, 1813, Minute Book, 1813–1818, Southern Historical Collection, M1-4485.

158. Prince William County Deed Book 5: 270, 283; 10: 261–264; TFP, 5: 604

159. See Occoquan Hotel Accounts, John Tayloe Account Book, Tayloe Family Papers, VHS, Mss1T2118C6.

160. Prince William County Deed Book 12: 273.

161. See, for example, Postal Ledger, 1812–1813, TFP, 3: 639–657.

162. Gideon Granger, General Post Office, City of Washington, to John Tayloe, Esq., Post-master at Richmond C.H., VA, TFP, 5:321.

163. Daybook, Stages Accounts (C4); Stages Accounts, John Tayloe Account Book, Tayloe Family Papers, VHS, Mss1T2118C6.

164. John Tayloe, Washington, to The Hon[orable] Post Mast[er] Gen[era]l, October 1, 1819 and October 2, 1819, John Tayloe Letter Book, Tayloe Family Papers, VHS, Mss1T2118C1.

165. Stages Account, John Tayloe Account Book.

166. Articles of Agreement, March 25, 1826, TFP, 5: 759; Prince William County Deed Book 10: 261–264.

167. John Tayloe, Washington, to Mr. P. Hewitt of Alex[andri]a, September 20, 1822; John Tayloe, Washington, to Hon[orable] Postmaster Gen[eral], October 10, 1822; William Henry Tayloe, Washington, to Hon[orable] Postmaster Gen[eral], October 10, 1822, in John Tayloe Letter Book (C1).

168. Botetourt County Deed Book 10: 47, 71; TFP, 5: 647. For a history of Cloverdale's earliest years, see Jeffrey C. Turner, "Cloverdale

Furnace: A Century of Iron Manufacture in Botetourt County, Virginia, 1789–1889," Master's Thesis, Virginia Polytechnic Institute, 1984; F. B. Kegley, *Kegley's Virginia Frontier: The Beginning of the Southwest, the Roanoke of Colonial Days, 1740–1783* (Roanoke, Va., 1938), 496, 511; Helen Prillaman, *Places Near the Mountain: From the Community of Amsterdam, Virginia up the Road to Catawba* (Roanoke, Va., 1985), 150–162.

169. Advertisement by William Gordon on behalf of John Tayloe, Beale Family Papers, VHS, Mss1B3658A16; Botetourt County Deed Book 10: 152; 11: 481, 550; Tayloe's Slitting Mill Forge and Brunswick Forge Logs, Tate Thompson Brady Papers.

APPENDICES

Appendix 1
Tayloe Family Genealogy

William Tayloe (1645-1710)
M
Anne Corbin

- Elizabeth Tayloe (1686) M Griffin Fauntleroy
- John Tayloe (1687-1747) M Elizabeth Gwynn Lyde
- William Tayloe (1694-1770) M Letitia Wormeley?

Children of John Tayloe and Elizabeth Gwynn Lyde:

- William Tayloe (died young)
- John Tayloe II (1721-1779) M Rebecca Plater
- Betty Tayloe (1721-1784) M Richard Corbin
- Anne Corbin Tayloe (1723-1784) M Mann Page II

Children of John Tayloe II and Rebecca Plater:

- Elizabeth (**1750-1825**) M Edward Lloyd IV
- Rebecca (1752-1797) M Francis Lightfoot Lee
- Anne Corbin (**1753-1835**) M Thomas Lomax
- Eleanor (1756-1815) M Ralph Wormeley V
- Mary (**1759-1835**) M Mann Page III
- Catherine (1761-1798) M Landon Carter II
- Sarah (1765-1834) M William Washington
- John III (1771-1828) M Anne Ogle
- Jane (1774-1816) M Robert Beverley

Children of John III and Anne Ogle:

- John IV (1793-1824) M Maria Forrest
- Henrietta (1794-1832) M H.C.S. Key
- Benjamin Ogle (1796-1868) M Julia Dickinson (1) Phoebe Warren (2)
- Rebecca Plater (1797-1815) (unmarried)
- William Henry (1799-1871) M Henrietta Ogle
- Ann (1800) (died young)
- Catherine (1801-?) M James Baker
- Edward Thornton (1803-1876) M Mary Ogle
- George Plater (1804-1897) M Mary Elizabeth Langhorne
- Henry Augustine (1808-1903) M Narcissa Jamieson
- Elizabeth Merry (1806-1832) M Robert W. Carter II
- Charles (1810-1847) M Virginia Turner
- Virginia (1813-1883) (unmarried)
- Anne Ogle (1814-1876) M Henry Howell Lewis
- Lloyd (1815-1816)

Appendix 2
Tayloe Family Land Transactions, Prince William County, 1731–1831

Transaction	*Date*	*With*	*Acreage*	*Tract Name*	*Location*	*Cost*
Purchase	October 8, 1731	Benjamin Grayson	2,367	Neabsco		
Purchase	March 20, 1732	Francis Awbrey	4,700	Neabsco	Truro Parish	£60 sterling
Purchase	October 18, 1732	Benjamin Grayson	1,214	Neabsco	Goose Creek?	
Purchase	April 14, 1733	John Goffin	200	Neabsco		
Purchase	April 15, 1733	John Gosling	200	Neabsco	On Neabsco Run	£20 sterling
Purchase	April 22, 1734	Abraham and William Farrow	375	Neabsco	Between the Chapman and Neabsco Creek	£80 Virginia currency
Purchase	May 20, 1735	Peter and Charles Cornwell	100	Neabsco		
Purchase	July 15, 1739	Francis Awbrey	3,242	Kittockton (Ketockton)	Becomes Fairfax County	
Sale	March 21, 1740	Willoughby Newton	700		On Rorksy Cedar Run	£150 sterling
Purchase	July 28, 1740	James Maxwell	500	Neabsco		£65 Virginia currency
Sale	September 28, 1749	George Calvert	150	Neabsco		
Sale	November 5, 1753	Tom Keys	254	Neabsco		
Sale	November 17, 1753	Jacob Calvert	180	Neabsco		
Purchase	March 25, 1755	Peyton	79	Occoquan		£150 Virginia currency
Purchase	May 24, 1755	Blackburn	1,715	Occoquan		£300 Virginia currency
Sale	November 9, 1755	B. Ewell	500	Neabsco		
Purchase	December 26, 1755	Richard Kenner	200	Neabsco	North side Neabsco Creek (improvements)	£150 Virginia currency
Purchase	July 4, 1760	William Fairfax estate	296	Occoquan	Marumsco Creek (improvements)	£44 Virginia currency

Purchase	1760	John Muschett	1,014	Occoquan	South side Occoquan Run	£200 Virginia currency
Purchase	1763?	Jones	2,146	Neabsco	On Quantico Creek	
Purchase	January 21, 1763	John Ballendine	1,070	Occoquan	South side Occoquan River (warehouse, forge, big house)	£150 Virginia currency
Sale	September 18, 1763	Jabes Downman	296	Occoquan	Marumsco Creek	£65.15 Virginia currency
Purchase	October 7, 1763	Hector Ross	457	Occoquan	On branch of Occoquan River	£100 Virginia currency
Purchase	1764	Singleton	600	Occoquan		£100 Virginia currency
Rental	June 6, 1764	John Hancock			"marsh ground" on Neabsco Creek, for 21 years	£5 Virginia currency per year
Purchase	August 31, 1764	Benjamin Grayson	263	Occoquan	Dettinger Parish	£210 Virginia currency
Purchase	November 3, 1764	Francis, George, and Daniel Peake	300	Occoquan		£65 Virginia currency
Purchase	1765	Henry Downy	400	Occoquan		£1200 sterling
Purchase	1765	Charles Croxall	33.5	Occoquan		£600 sterling
Purchase	October 7, 1765	Ashton	457	Occoquan		£100 Virginia currency
Purchase	October 8, 1765	John Hancock	500	Neabsco	Between mouth of the Occoquan River and Neabsco Creek	£1000 Virginia currency
Purchase	May 7, 1766	James Douglass (Ballendine)	300	Occoquan	(storehouse, mills, bake oven, forge)	£578 Virginia currency
Purchase	1767	Harrison	100	Occoquan		£200 Virginia currency
Purchase	April 20, 1767	Simon Hancock	100	Neabsco		£50 Virginia currency
Purchase	May 3, 1767	Jabez Downman	112	Occoquan	On Marumsco Creek and Occoquan River	

Purchase	October 28, 1768	John Linton	370	Deep Hole	Adjacent to Occoquan and Potomac Rivers	£857.10 Virginia currency
Purchase	June 27, 1769	Northern Neck Proprietor	880	Occoquan	Adjacent other Tayloe land on branch of Occoquan River	
Purchase	August 26, 1773	John Hancock	373	Deep Hole?		
Purchase	August 9, 1774	Robert Moseley	350	Neabsco		
Purchase	January 21, 1793	John Tayloe Thornton	724			£1458 Virginia currency
Sale	November 26, 1793	John Gibson	724			£1458 Virginia currency
Sale	September 29, 1794	John Sowden	227	Neabsco	North side Quantico Run	£255.13 Virginia currency
Sale	October 9, 1795	Archibald Carr	350	Neabsco	North side Quantico Run	£350 Virginia currency
Sale	October 9, 1795	Daniel Carr	350	Neabsco	South side Quantico Run	£350 Virginia currency
Sale	October 9, 1795	John Cannon	100	Neabsco		£80 Virginia currency
Sale	October 9, 1795	Luke and Charles Cannon	300	Neabsco	On Quantico Run	£300 Virginia currency
Sale	June 6, 1795	M. Whilson	79			
Sale	June 7, 1796	Isaac McPherson and Nate Ellicott	79	Occoquan	Wharf extending into the Occoquan River	£400 Virginia currency
Exchange	January 1, 1798	Presley Thornton		Occoquan	Receives Presley Thornton's ¼ interest in Occoquan Ironworks	In exchange for P.T. being released from all debts related to the works
Sale	March 29, 1800	Philip Davis	123	Occoquan	On Occoquan Run	£233.14 Virginia currency

Purchase	November 8, 1800	Benjamin Hooe	200	Occoquan	On the bay of Occoquan	£1500 Virginia cur-
Sale	November 30, 1801	Alexander Henderson	78.75	Neabsco	On west side of main road leading from Dumfries	£100 Virginia currency
Purchase	September 11, 1805	William Duvall	133	Neabsco	Includes two roads	
Sale	February 6, 1809	John Linton	126	Neabsco	On north branch of Quantico run	$630.81
Sale	November 26, 1813	Williams	98			
Purchase	March 15, 1814	John Howe	1 lot	Occoquan	#21?	
Purchase	August 6, 1814	Robert Walsh	2 lots	Occoquan	#45, 46	
Sale	December 23, 1825	Michael Cleary	2 lots	Occoquan	Lots #45 and 46	$1
Sale	December 23, 1825	Michael Cleary	1 lot	Occoquan	Lot #21 on Mill and Ellicott Streets	$250.00
Sale	January 14, 1826	Richard Slade	1,000	Neabsco	Hutts Tract, part of Neabsco	
Sale	December 1826	John Bramill	135	Neabsco	On stage road between Dumfries and Occoquan	$337.50
Sale	December 9, 1826	Arrington Evans	600	Neabsco		$1,800.00
Sale	December 14, 1826	Sampson Windsor	320	Neabsco		$640.00
Sale	November 17, 1827	Chapman Clark	100		On the Occoquan Road	$300.00
Sale	December 11, 1827	Arrington Evans	713	Neabsco		$2,140.25
Sale	December 11, 1827	Sampson Windsor	447	Neabsco		$894.07
Sale	November 17, 1828	Richard Slade	561	Neabsco		$1,176.23
Sale	October 22, 1829	Leonard Love	121	Neabsco		$302.76
Sale	October 22, 1829	Richard Stonnel	149	Neabsco		$373.83
Sale	April 11, 1831	John Weedon	165	Neabsco		$413.12
Sale	April 11, 1831	Newman Davis	100	Neabsco	On Neabsco water	$252.37
Sale	April 11, 1831	Araminta and Zebulon Kankey	126	Neabsco Hotel		$800.00

Appendix 3
The Chesapeake Iron Industry, Known Furnaces and Forges, ca. 1700–1830[a]

Ironworks Name	*Furnace, Forge, Both?*	*County*	*Owner(s)*	*Dates of Operation*	*# Hands*	*Annual Production*
Falling Creek	Furnace	Chesterfield, VA		1619–1622		
Tubal	Furnace	Spotsylvania, VA	Alexander Spotswood	1714–1780?		
Principio	Both (4 fu, 2 fo)	MD & VA	British merchants	1720–1760?	137	2458 tons pig (1736)
Bristol	Furnace	Westmoreland, VA	Bristol merchants, John Tayloe I	1721–1742		
Pine	Forge	Shenandoah, VA		1725–1835	47	170 tons/yr (1820)
Accokeek	Furnace	Stafford, VA	Principio Company	1726–1753	26	500 tons/yr
Massaponax Air	Furnace	Spotsylvania, VA	Alexander Spotswood	1727–1760		
Fredericksville	Furnace	Spotsylvania, VA	Charles Chiswell, Alex Spotswood	1728–1760?	100	
Baltimore	Both (2, 2)	Baltimore, MD	D. Dulany, Carrolls, B. Tasker	1731–1770?	150	376 tons (1750)
Germanna	Furnace	Spotsylvania, VA	Alexander Spotswood	1732–		
Snowden	Both	Anne Arundel, MD	John Snowden	1734–1776		
Neabsco	Furnace	Prince William, VA	John Tayloe I, II	1737–1810?	40	
Stagland	Forge	Richmond City, VA	William Byrd II	1740?–1800		
Vestal	Forge	Jefferson, VA	Thomas Maybury	1742–		
Kingsbury	Furnace	MD		1744–		
Bush River	Furnace	MD	Webster, Hall, Giles	1746–		
Nottingham	Both	Baltimore, Harford, MD		1746–1765	153	
Concord	Forge	Spotsylvania, VA	Benjamin Gryme, Wm Champe	1750–1771		

Cumberland	Forge	Harford, MD		1750?–		
Recovery	Furnace	Spotsylvania, VA	Benjamin Gryme, A. Thornton	1750–1771		
Rock	Forge	Harford, MD	George Rock	1750?–		
Providence	Forge	New Kent, VA	William Holt	1755?–1784?		
Elk Ridge	Furnace	Howard, MD	Dorsey	1756–		
Mossy Creek	Both	Augusta, VA	John Miller	1757–1767		"600 tons/yr" (TJefferson)
Marley	Furnace	Harford, MD	Dorsey	1759–		
Northampton	Both	Baltimore, MD	Charles Ridgely	1759–1829		
Occoquan	Forge	Prince William, VA	Tayloe II, Thornton, Ballendine	1759–1780?	50	
Occoquan	Furnace	Prince William, VA	Tayloe II, Thornton, Ballendine	1755–1782	50	
Elk	Forge	Cecil, MD		1760–1781	16	
Grant's	Furnace	Rockbridge, VA		1760?–		
Gryme's	Forge	Spotsylvania, VA	Benjamin Gryme	1760?–		
Hampton	Furnace	Emmitsburg, MD		1760–		
Hockley	Forge	Harford, MD	Dorsey	1760–		
Isabella (Redwell)	Furnace	Page, VA	1760–1841	150		220 tons, 300 ton castgs/yr (1820)
Marlboro	Forge	Frederick, VA	Isaac Zane?	1760–1828		"150 tons bar/wk" (TJefferson)
Mossy Creek	Furnace	Augusta, VA	Henry Miller	1760–1841		
Zane's	Furnace	Frederick, VA	Isaac Zane	1760?–1845?		

Appendix 3 (*continued*)
The Chesapeake Iron Industry, Known Furnaces and Forges, ca. 1700–1830[a]

Ironworks Name	*Furnace, Forge, Both?*	*County*	*Owner(s)*	*Dates of Operation*	*# Hands*	*Annual Production*
Onion	Both (1 fu, 2 fo)	Joppa, MD	Stephen Onion	1760?–1770?		
Unicorn	Forge	Queen Anne's, MD		1760?–1770?		
Moore's	Forge	King and Queen, VA	Bernard Moore	1761–1781	53	
Potts-Wilson	Forge	Fairfax, VA	John Semple, Potts, Wilson	1763–1828		
Hampton	Furnace	Emmitsburg, MD	Carroll	1764–1770?		
Legh	Furnace	Carroll, MD		1765–1770?		
Nottingham	Furnace	Baltimore, MD	Charles Ridgely	1765–		
Antietam	Forge	Washington, MD	Samuel Hughes	1766–1865		
Frederick	Forge	Washington, MD	Samuel Hughes	1766–		
Antietam	Furnace	Washington, MD	Samuel Hughes	1768–1865		
Green Spring	Furnace	Washington, MD	James Johnson & Co.	1768–1779?		
Jacques	Forge	Washington, MD		1770–		
Legh	Furnace	Frederick, VA	Legh Master	1770?–		
Albemarle	Both	Albemarle, VA	John Old	1771–1796		
Mt. Aetna	Furnace	Washington, MD	Samuel Hughes	1771–1850?		
Old's	Forge	Albemarle, VA	John Old	1771–1796		
Marlboro	Furnace	Frederick, VA	Isaac Zane?	1772–1828		"600 tons/yr" (TJefferson)

Washington	Both	Franklin, VA	Donelson, Calloway, Early	1774–1850	200	
Catoctin	Furnace	Catoctin, MD	James Johnson & Co.	1774–1900?		12–18 tons/wk
Bush Creek	Forge	Catoctin, MD	James Johnson & Co.	1774–		
Bear Garden	Furnace	Buckingham, VA	John Ballendine?	1775–1840		"1000 tons/yr" (TJefferson)
Maho (Fawcett)	Furnace	Rockingham, VA	Benjamin Fawcett, Benj Savage	1775–1816		
Oxford	Furnace	Campbell, VA	David & Frederick Ross	1775–1837	500	"1600 tons/yr" (TJefferson)
Rappahannock	Forge	Stafford, VA	James Hunter	1775?		"300 tons bar/yr" (TJefferson)
Spring	Forge	Rockingham, VA		1775–		
Westham	Forge	Richmond City, VA	John Ballendine	1776–		
Calloway's	Forge	Franklin, VA	Calloway	1779–		"150 tons bar/yr" (TJefferson)
Calloway's	Furnace	Campbell, VA	J. Calloway, David Ross	1779–1837		"600 tons/yr" (TJefferson)
Dougherty	Forge	Rockbridge, VA		1779?–1840		
Broce's	Forge	Montgomery, VA		1780–		200 lbs/day
Marshall	Furnace	Rockingham, VA		1780?–		
Speedwell Providence	Furnace	Wythe, VA		1781?–1866		
Shelor's	Furnace	Floyd, VA	Daniel Shelor	1785–1815		
Brunswick	Forge	Botetourt, VA	John Tayloe III	1786–1851		145 tons bar/yr (1820)

Appendix 3 (*continued*)
The Chesapeake Iron Industry, Known Furnaces and Forges, ca. 1700–1830[a]

Ironworks Name	*Furnace, Forge, Both?*	*County*	*Owner(s)*	*Dates of Operation*	*# Hands*	*Annual Production*
Cloverdale	Furnace	Botetourt, VA	Robt Harvey, Beverley, Tayloe III	1787–1849	124	600 tons bar, 30 tons cast/yr (1820)
Harvey's	Forge	Botetourt, VA	Robert Harvey	1787?–1825		100 tons bar/yr (1820)
Martha	Furnace	Botetourt, VA	Robert Harvey	1787?–1825		230 tns bar, 20 tns hware/yr (1820)
Speedwell	Both	Roanoke, VA	Robert Harvey	1787–1825		
Keep Tryst	Furnace	Jefferson, WV	John Potts, William Wilson	1788?–		
Stonewall	Furnace	Appomattox, VA	David Ross?	1788–1845		
McClure	Forge	Rockbridge, VA	McClure	1789?–1820	16	
Moore's	Furnace	Rockbridge, VA		1789?–1829		
Carron	Furnace	Franklin, VA		1790–1860		
David Ross	Forge	Hawkins, TN	David Ross	1790–1864		
David Ross	Furnace	Sullivan, TN	David Ross	1790–		
Elk	Forge	Franklin, VA		1792–1820?		
Etna	Furnace	Botetourt, VA	Wilson, Maybury, Wm Weaver	1792–1859		1100 tons, 20 tons hware/yr (1820)
Point Hope	Both	Grayson, VA	Matthew Dickey	1792?–		
Wythe	Furnace	Wythe, VA	James Byrne, William Love	1792–1890		
Harvey's	Forge	Franklin, VA	Robert Harvey	1795–		
Williams'	Forge	Bath, VA	Elisha Williams	1795–	3	

Slitting Mill	Forge	Botetourt, VA	John Tayloe III	1797–1851		
Poplar Camp	Furnace	Wythe, VA	David Pierce	1798–1820s		
Union (Paoli)	Forge	Shenandoah, VA	Blackford & Arthur	1799–1850	45	
Beverley	Furnace	Rockbridge, VA	McClure	1800?–1818		
Buffalo	Forge	Rockbridge, VA	William Weaver	1800–1868	20	150 tons bar/yr (1850)
Gibraltar	Forge	Rockbridge, VA	T. Maybury, S. Jordan, Davis	1800–1845		
Graham's	Forge	Wythe, VA	David Graham	1800–1856	31	161 tons blooms, 23 tons bar/yr
John's Creek	Forge	Craig, VA		1800–1830		
Jordan's	Furnace	Rockbridge, VA	John Jordan	1800–1860?		
Lagrange	Furnace	Appomatox, VA		1800–1843		
Lydia	Furnace	Rockbridge, VA	William Moore, William Weaver	1800–1850		
McClure's	Foundry	Rockbridge, VA	McClure	1800?–1818		
Mount Torry	Furnace	Augusta, VA		1800–1885		
Port Republic	Forge	Rockingham, VA	John Carthrea	1800–1870s		
Retreat	Furnace	Botetourt, VA	William Wilson, Thomas Maybury	1800–1815		
Rumsey	Furnace	Alleghany, VA		1800?–1854		
Union B	Forge	Augusta, VA		1800–1850		
Tilthammer	Forge	Bath, VA	Elisha Williams	1800–1840		
Cranberry	Furnace	Carroll, VA		1801–1840		
Columbia	Furnace	Shenandoah, VA		1803–1823?	115	560 tons/yr (1820)
Shenandoah	Both	Page, VA (Elkton, WV)		1805–1863		

Appendix 3 (*continued*)

The Chesapeake Iron Industry, Known Furnaces and Forges, ca. 1700–1830[a]

Ironworks Name	*Furnace, Forge, Both?*	*County*	*Owner(s)*	*Dates of Operation*	*# Hands*	*Annual Production*
Catawba	Furnace	Botetourt, VA		1807–1849		
Bellona	Foundry	Powhatan, VA	John Clarke	1810–1877		
Mount Vernon	Forge	Rockingham, VA		1810–1855		150 tons/yr
Raven Cliff	Furnace	Wythe, VA	Joseph Bell	1810–1856		
Speedwell #1	Forge	Page, VA		1811–		
Kennedy (Canada)	Furnace	Augusta, VA		1812–1820		
Tilthammer	Forge	Franklin, VA	Townes	1814–1819		
Garber	Foundry	Augusta, VA	Albert Garber	1815–	5	65 tons castings/yr
Speedwell #2	Forge	Page, VA		1815–		
Fine Creek	Forge	Powhatan, VA	McRae, Brooks, & Archer	1816–1823		
Cumberland Gap	Furnace	Lee, VA		1819–1873		800 tons/35 wks
Rebecca	Furnace	Botetourt, VA	William Ross	1819–1863		
Beechenbrook	Foundry	Rockbridge, VA	John Jordan	1820–1890		
Blue Spring	Furnace	Alleghany, VA	John Jordan	1820–1840		
Henrietta	Furnace	Shenandoah, VA		1820–1864		
Liberty	Furnace	Shenandoah, VA		1821–1905		387 tons/20 wks
Capon	Furnace	Hardy, WV		1822–		300 tons/16 wks

Clifton	Forge	Alleghany, VA	John Jordan	1824–	30	150 tons bar/yr
Elk Creek	Furnace	Nelson, VA		1824–1850		
Brown's	Forge	Washington, VA	James Brown	1825–1841	15	2.5 tons bar/yr
James River	Forge	Botetourt, VA	William Ross	1825–1848		
Lebanon Valley	Forge	Rockbridge, VA		1825–		
Lime Rock	Furnace	Litchfield, VA		1825–		205 tons/22 wks (1856)
Bath	Forge	Rockbridge, VA	William Weaver	1827–1850	60	160 tons bar/yr (1829)
Lucy Selina	Furnace	Alleghany, VA	Ira Jordan	1827–1852	70	1250 tons, 25 tons castgs/yr (1850)
Liberty	Forge	Shenandoah, VA	Walter Newman	1828–		

[a]Information for Appendix 3 comes from the following sources: Iron Industry Reports, Tate Thompson Brady Papers, Virginia Historical Society; Tayloe Family Papers, VHS; *Virginia Gazette*; *Maryland Gazette*; William Peden, ed., *Notes on the State of Virginia* by Thomas Jefferson (New York, 1982); Michael Robbins, *The Principio Company: Iron-Making in Colonial Maryland, 1720–1781* (New York, 1986); David Curtis Skaggs, "John Semple and the Development of the Potomac River Valley, 1750–1773, *Virginia Magazine of History and Biography* 92 (3) (1984), 287; William Reynolds, "An Account of the Albemarle Iron Works," *The Magazine of Albemarle County History* 50 (1992), 39–57; Arthur Cecil Bining, *British Regulation of the Colonial Iron Industry* (Clifton, N.J., 1973); Keach Johnson, "Genesis of the Baltimore Company," *Journal of Southern History* 19 (May 1953), 157–179; Frances C. Robb, "Industry in the Potomac River Valley, 1760–1860," unpublished dissertation, West Virginia University, 1991; Ronald Hoffman, *A Spirit of Dissension: Economics, Politics, and the Revolution in Maryland* (Baltimore, 1973); Edward Heite, "The Pioneer Phase of the Chesapeake Iron Industry: Naturalization of a Technology," *Quarterly Bulletin of the Archaeological Society of Virginia* 38 (3) (1983), 133–181.

Bibliography

Archival Primary Sources:

Colonial Williamsburg Foundation Research Library:
Inventory of Estates of John Tayloe and Presley Thornton (Occoquan)

Library of Virginia (LVA):
John Tayloe Ledger, 1747–1787, Essex County Business Records

Maryland State Archives (MSA):
Baltimore County Certificates, Patented, Folder 4818
Land Records—Anne Arundel County, Baltimore County

Virginia City and County Records (also located at the Library of Virginia):
Fredericksburg Court Records, File 171
Loudoun County Records
Prince William County Records
Richmond County Records

Virginia Historical Society (VHS):
Tate Thompson Brady Papers
William Byrd Letterbook
Carter Family Papers
Lee Family Papers
Tayloe Family Papers
John Tayloe Papers
John Wadman Petition

Published Primary Sources:

Hening, William Waller, ed. *The Statutes at Large: Being a Collection of All the Laws of Virginia, from the First Session of the Legislature, in the Year 1619.* Charlottesville: University Press of Virginia, 1969.

Joyner, Peggy Shomo, comp. *Abstracts of Virginia's Northern Neck Warrants & Surveys, 1710–1780.* 3 Vols. Portsmouth, Virginia: Peggy Joyner, 1986.

McIlwaine, H. R., ed. *Executive Journals of the Council of Colonial Virginia.* Richmond: Virginia State Library, 1925.

———. *Minutes of the Council and General Court of Colonial Virginia.* Richmond: Virginia State Library, 1979.

Mair, John. *Book-Keeping Moderniz'd; or, A Methodical Treatise of Merchant-Accompts; or, Merchant-Accompts by Double Entry.* Edinburgh, 1736, 1773.

Palmer, William P., ed. *Calendar of Virginia State Papers and Other Manuscripts.* New York: Kraus Reprint Corp., 1968.

Peden, William, ed. *Notes on the State of Virginia,* by Thomas Jefferson. New York: W. W. Norton & Co., 1954, 1982.

Virginia Land Records: From the Virginia Magazine of History and Biography, the William and Mary Quarterly, and Tyler's Quarterly. Baltimore: Genealogical Publishing Co., Inc., 1982.

"Virginia Militias in the Revolution," *Virginia Magazine of History and Biography,* 15 (June 1908): 87–92.

Vogt, John and T. William Kethley, Jr., comps. *Stafford County, Virginia, Tithables: Quit Rents, Personal Property Taxes, and Related Lists and Petitions, 1723–1790,* 2 vols. Athens, GA: Iberian Publishing Co., 1990.

Wright, Louis B., ed. *Letters of Robert Carter, 1720–1727: The Commercial Interests of a Virginia Gentleman.* San Marino, CA: The Huntington Library, 1940.

Secondary Sources:

Baxter, W. T. "Accounting in Colonial America," in A. C. Littleton and B. S. Yamey, eds., *Studies in the History of Accounting.* Homewood, Illinois: R. D. Irwin, 1956, 272–287.

Berry, Thomas, "The Rise of Flour Milling in Richmond," *The Virginia Magazine of History and Biography,* 18 (October 1970): 387–408.

Bezís-Selfa, John. "Slavery and the Disciplining of Free Labor in the Colonial Mid-Atlantic Iron Industry," *Pennsylvania History,* 64 (Summer 1997): 270–286.

Bining, Arthur Cecil. *British Regulation of the Colonial Iron Industry.* Clifton, NJ: A. M. Kelley, 1973.

Bradford, Samuel Sydney. "The Negro Ironworker in Ante-Bellum Manufacturing," *Journal of Southern History,* 25 (May 1959): 201–204.

Braun, Rudolf. "The Rise of a Rural Class of Entrepreneurs," *Journal of World History,* 10 (1967): 551–566.

Breen, T. H. *Tobacco Culture: The Mentality of the Great Tidewater Planters on the Eve of the Revolution.* Princeton: Princeton University Press, 1985.

Bruce, Kathleen. *Virginia Iron Manufacturers in the Slave Era.* New York: A. M. Kelley, 1968.

Brydon, G. McClaren. "The Bristol Iron Works in King George County," *Virginia Magazine of History and Biography,* 42 (2) (April 1934): 97–102.

Chaplin, Joyce E. *An Anxious Pursuit: Agricultural Innovation and Modernity in the Lower South, 1730–1815.* Chapel Hill: Published for the Institute of Early American History and Culture by the University of North Carolina Press, 1993.

Clark, Christopher. *The Roots of Rural Capitalism: Western Massachusetts, 1780–1860.* Ithaca, New York: Cornell University Press, 1990.

Clemens, Paul G. E. "The Operation of an Eighteenth-Century Chesapeake Tobacco Plantation," *Agricultural History,* 44 (July 1975): 517–531.

Clowse, Converse D. "Shipowning and Shipbuilding in Colonial South Carolina: An Overview," *American Neptune,* 44 (4) (1984): 221–244.

Courts, Frank. "A 'Quiet' Legacy: The Tayloes of Virginia," *Northern Neck of Virginia Historical Magazine,* 42(1) (1992): 4851–4865.

Daniels, Christine. "Gresham's Laws: Labor Management on an Early-Eighteenth-Century Chesapeake Plantation," *Journal of Southern History,* 62 (May 1996): 205–238.

———. "'WANTED: A Blacksmith Who Understands Plantation Work': Artisans in Maryland, 1700–1810," *William and Mary Quarterly,* 3d. ser., 50 (October 1993): 743–767.

Dew, Charles B. *Bond of Iron: Master and Slave at Buffalo Forge.* New York: W. W. Norton & Company, 1994.

———. "David Ross and the Oxford Iron Works: A Study of Industrial Slavery in the Early Nineteenth-Century South," *William and Mary Quarterly,* 3d ser., 31 (April 1974): 189–224.

Dunn, Richard S. "Annual Work Rhythms at Mount Airy," unpublished paper presented at the Re-creating the World of the Virginia Plantation conference, University of Virginia, June 1990.

———. "A Tale of Two Plantations: Slave Life at Mesopotamia in Jamaica and Mount Airy in Virginia, 1799–1828," *William and Mary Quarterly,* 3d ser., 36 (January 1977): 32–65.

———. "Two Slave Women: Sarah Affir of Jamaica and Winney Grimshaw of Virginia," in *Lois Green Carr: The Chesapeake and Beyond—A Celebration.* Crownsville, MD: Maryland Historical and Cultural Publications, 1992, 37–53.

Earle, Carville and Ronald Hoffman. "Urban Development in the Eighteenth-Century South," *Perspectives in American History,* 10 (1976): 7–80.

Egnal, Marc. "The Economic Development of the Thirteen Continental Colonies, 1720 to 1775," *William and Mary Quarterly,* 3d. Ser., 32 (2) (April 1975): 191–222.

Ekirch, A. Roger. *Bound for America: The Transportation of British Convicts to the Colonies, 1718–1775.* New York: Oxford University Press, 1987.

Frey, Sylvia R. *Water from the Rock: Black Resistance in a Revolutionary Age.* Princeton: Princeton University Press, 1991.

Gallay, Alan. *The Formation of a Planter Elite: Jonathan Bryan and the Southern Colonial Frontier.* Athens: University of Georgia Press, 1989.

Genovese, Eugene. *The Political Economy of Slavery: Studies in the Economy and Society of the Slave South.* New York: Random House, 1965.

——— and Elizabeth Fox-Genovese. "The Slave Economies in Political Perspective," *Journal of American History,* LXVI (1979): 7–23.

Gutman, Herbert G. *The Black Family in Slavery and Freedom, 1750–1925.* New York: Vintage Books, 1976.

Harrison, Fairfax. *Landmarks of Old Prince William: A Study of Origins in Northern Virginia.* 2 vols. Baltimore: Gateway Press, 1987.

Harper, Robert R. *Richmond County, Virginia, 1692–1992: A Tricentennial Portrait.* Alexandria, VA: O'Donnell Publications, 1992.

Haywood, C. Robert. "Economic Sanctions: Use of the Threat of Manufacturing by the Southern Colonies," *Journal of Southern History,* 25 (2) (May 1959): 207–219.

Heite, Edward. "The Pioneer Phase of the Chesapeake Iron Industry: Naturalization of a Technology," *Quarterly Bulletin of the Archaeological Society of Virginia,* 38(3) (1983): 133–181.

Henretta, James A. "Families and Farms: *Mentalité* in Pre-Industrial America," *William and Mary Quarterly,* 3d ser., 35 (1978): 3–32.

Hoffman, Ronald, ed. *The Economy of Early America: The Revolutionary Period, 1763–1790.* Charlottesville, Virginia: Published for the United States Capitol Historical Society by the University Press of Virginia, 1988.

Hudgins, Carter L. and Douglas W. Sanford. "Neabsco Mills Ironworks Site, National Register of Historic Places Registration Form," Center for Historical Preservation, Mary Washington College, 1993.

Isaac, Rhys. *The Transformation of Virginia, 1740–1790.* New York: W. W. Norton & Company, 1982.

Johnson, Keach. "The Baltimore Company Seeks English Markets: A Study of the Anglo-American Iron Trade, 1731–1755," *William and Mary Quarterly,* 3d ser., 16 (1959): 37–60.

———. "Genesis of the Baltimore Company," *Journal of Southern History,* 19 (May 1953): 157–179.

Kelso, William M. "Shipbuilding in Virginia, 1763–1774," *Records of the Columbia Historical Society,* XLVIII (1971): 1–13.

Klingaman, David. "Development of the Coastwise Trade of Virginia in the Late Colonial Period," *Virginia Magazine of History and Biography,* 77 (1969): 25–45.

———. "The Significance of Grain in the Development of the Tobacco Colonies," *Journal of Economic History,* XXIX (1969): 268–278.

Kuethe, J. Louis. "A List of Maryland Mills, Taverns, Forges, and Furnaces of 1794," *Maryland Historical Magazine,* 21 (June 1936): 155–169.

Kulikoff, Allan. *Tobacco and Slaves: The Development of Southern Cultures in the Chesapeake, 1680–1800.* Chapel Hill: Published for the Institute of Early American History and Culture by the University of North Carolina Press, 1986.

Land, Aubrey C. *Bases of the Plantation Society.* Columbia, South Carolina: University of South Carolina Press, 1969.

———. "Economic Base and Social Structure: The Northern Chesapeake in the Eighteenth Century," *Journal of Economic History,* 25 (1965): 639–654.

Lewis, Ronald. *Coal, Iron, and Slaves: Industrial Slavery in Maryland and Virginia, 1715–1865.* Westport, Connecticut: Greenwood Press, 1979.

McCusker, John J. and Russell R. Menard. *The Economy of British America, 1607–1789.* Chapel Hill: Published for the Institute of Early American History and Culture by the University of North Carolina Press, 1985.

Main, Jackson Turner. "The One Hundred," *William and Mary Quarterly,* 3d ser., 11 (1954): 354–384.

Middleton, Arthur Pierce. "Ships and Shipbuilding in the Chesapeake Bay and Tributaries," in Ernest M. Eller, ed., *The Chesapeake Bay in the American Revolution.* Centerville, MD: 1981, 98–132.

———. *Tobacco Coast: A Maritime History of the Chesapeake Bay in the Colonial Era.* 2 Vols. Newport News, VA: Mariners Museum, 1953.

Morton, Louis. *Robert Carter of Nomini Hall: A Virginia Planter of the Eighteenth Century.* Charlottesville, Virginia: Dominion Books, 1965.

Mullin, Gerald W. *Flight and Rebellion: Slave Resistance in Eighteenth-Century Virginia.* London: Oxford University Press, 1972.

O'Dell, Jeffrey M. "The War of Independence and the Northern Neck of Virginia," *Northern Neck of Virginia Historical Magazine,* 21(1) (December 1971): 2116–2176.

Papenfuse, Jr., Edward C. "Planter Behavior and Economic Opportunity in a Staple Economy," *Agricultural History,* 46 (April 1972): 297–312.

Paskoff, Paul. "Labor Productivity and Managerial Efficiency against a Static Technology: The Pennsylvania Iron Industry, 1750–1800," *Journal of Economic History,* XL (1980): 129–135.

Perkins, Edwin. "The Entrepreneurial Spirit in Colonial America: The Foundations of Modern Business History," *Business History Review,* 63 (1) (Spring 1989): 160–186.

Reynolds, William. "An Account of the Albemarle Iron Works," *The Magazine of Albemarle County History,* 50 (1992): 39–57.

Robb, Frances C. "Industry in the Potomac River Valley, 1760–1860," unpublished dissertation, West Virginia University, 1991.

Robbins, Michael W. *The Principio Company: Iron-Making in Colonial Maryland, 1720–1781.* New York: Garland Publishers, 1986.

Sanford, Douglas W. "The Neabsco Mills Ironworks: A Phase I and Phase II Archaeological Survey, 1993," unpublished manuscript on file with the Virginia Department of Historic Resources, Richmond, Virginia.

Skaggs, David Curtis. "John Semple and the Development of the Potomac River Valley, 1750–1773," *Virginia Magazine of History and Biography,* 92(3) (1984): 282–308.

Starobin, Robert S. *Industrial Slavery in the Old South.* New York: Oxford University Press, 1970.

Turner, Jeffrey C. "Cloverdale Furnace: A Century of Iron Manufacture in Botetourt County, 1789–1889," unpublished master's thesis, Virginia Polytechnic Institute, 1984.

Vickers, Daniel. "Competency and Competition: Economic Culture in Early America," *William and Mary Quarterly,* 3d ser., 47 (January 1990): 3–29.

Voke, Albert F. "Accounting Methods of Colonial Merchants in Virginia," *The Journal of Accountancy,* 41(7) (July 1926): 1–11.

Walsh, Lorena S. "Land, Landlord, and Leaseholder: Estate Management and Tenant Fortunes in Southern Maryland, 1640–1820," *Agricultural History,* 59 (1985): 373–396.

———. "Plantation Management in the Chesapeake, 1620–1820," *Journal of Economic History,* 49 (1989): 393–406.

———. "Summing the Parts: Implications for Estimating Chesapeake Output and Income Subregionally," *William and Mary Quarterly,* 3d. ser., 56 (1) (January 1999): 53–94.

Wilson, Robert J. *Early American Account Books: Interpretation, Cataloguing, and Use.* Nashville: University of Tennessee Press, 1981.

Author Biography

Laura Croghan Kamoie is a Visiting Assistant Professor of History at American University in Washington, D.C. She received her Ph.D. in American history from the College of William and Mary in 1999. This essay is based on her dissertation, "Three Generations of Planter-Businessmen: The Tayloes, Slave Labor, and Entrepreneurialism in Virginia, 1710–1830." Her areas of teaching and research interest include early American history, African-American history, and southern history. She is also a historical archaeologist and editor of the biannual magazine, *Washington History,* published by the Historical Society of Washington, D.C. She is married to Brian Kamoie and lives in Arlington, Virginia.